1986

INFERNO!

fourteen fiery tragedies of our time

HAL BUTLER

Henry Regnery Company • Chicago

Library of Congress Cataloging in Publication Data

Butler, Hal.
Inferno! Fourteen Fiery Tragedies of Our Time

1. Fires—North America. I. Title.
TH9448.B87 1975 973 75-11828
ISBN 0-8092-8352-2

To my wife, Eleanor, who understands
a writer's need for quiet and solitude
while working, and who graciously
sacrificed a woman's right to talk
so that this book could be written

Published by Henry Regnery Company
180 North Michigan Avenue, Chicago, Illinois 60601
Manufactured in the United States of America
Library of Congress Catalog Card Number: 75-11828
International Standard Book Number: 0-8092-8352-2

Published simultaneously in Canada by
Fitzhenry & Whiteside Limited
150 Lesmill Road
Don Mills, Ontario M3B 2T5
Canada

Contents

Introduction

Probably nothing in the world inspires more fear in man than uncontrolled fire. Although man, early in his long history, learned to harness this natural enemy and bend it to his own use, still it remains an implacable foe that will turn against him the moment he becomes careless or apathetic —often with disastrous consequences.

Fire was undoubtedly the first natural phenomenon man learned consciously to use for his own purposes. Primitive man's first introduction to fire was most likely in the form of a forest ignited by a bolt of lightning, and he undoubtedly fled in great horror before the resulting holocaust. It must have been thousands of years before man learned to make use of fire to warm himself and to cook his food, and thousands more before he discovered a way to make his own fire. Certainly it was not until many millenniums later, in the year 1827, that an English druggist named John Walker invented the first match. Today fire is man's ally in virtually everything he does—in heating his homes and buildings, preparing his food, manufacturing the many products used in the industrialized world. Fire even helped to launch man on what may become his greatest adventure—the exploration of space.

But fire is a fickle friend. It can also wipe out all of man's possessions in a matter of minutes. It can turn on its keeper like Frankenstein's monster, destroy everything he has built, and snuff out his life, if for just one moment man allows it to escape his control.

Inferno! is a book about famous—or infamous—fires. It traces the history and explores the reasons for some of the great fiery tragedies of our time. It is not intended to be a complete catalog of such disasters, but it does present the detailed stories of fourteen major fires that have occurred in North America between 1871 and 1958. It includes conflagrations that have consumed great cities and huge forests, as well as fire aboard ship, in a prison, at a theater, in a school, at a circus, in a hotel, at a nightclub, and in a factory—a potent warning that fire can occur wherever people congregate in large numbers. If the lessons to be learned from these tragic holocausts help prevent even one fire, this book will have served a purpose.

Hal Butler

1

The Great Chicago Fire (1871)

ROBERT Williams, chief fire marshall for the City of Chicago, was a frightened and frustrated man—frightened because he knew that Chicago was a tinderbox that could be ignited by a single carelessly tossed match, frustrated because nobody, other than himself, seemed to give a damn.

It was fall 1871. Chicago was in a period of rapid growth. From a village of four thousand people in 1840, it had grown to three hundred thousand by 1871. This growth demanded additional homes and places of business, and the mushrooming town threw up buildings quickly, aimlessly, carelessly to accommodate its rising population. By the year 1871 there were sixty thousand buildings spread over thirty-six square miles. Almost all of them were of wood. Even the newer buildings, made of stone or brick, had upper stories of wood with either tarred or wood-shingle roofs.

The homes of working-class families were particularly

vulnerable, and they stretched from downtown Chicago to the outskirts. Most of them were little more than shacks, and many of these flimsy shanties had wood piles for winter fuel stacked against their outer walls. Ramshackle barns and fragile sheds stood nearby. Shabby, weather-beaten tenements housed thousands of people in tiny cubicles. To make matters worse, all of these fire-prone structures were tied together by a network of wooden sidewalks and fences, making it an easy matter for fire to spread from one home to another.

There was no doubt that Chicago's growth and progress made it the midwest's most important city. But it was an endangered city, constantly teetering on the brink of disaster. Nobody knew this better than Fire Marshall Williams. Not only were the wooden buildings in the city highly combustible but the weather during the summer of 1871 had doubled the fire danger. From July 3 to the first part of October the city had received only two and a half inches of rain, rather than the eight or nine inches it was accustomed to. Every stick of wood in the city—in homes, business houses, factories—was as dry as tinder. And to make things worse, the merciless summer-long sun and lack of moisture had withered the leaves on the trees and dumped them in hills on the ground.

There wasn't much Williams could do about the problem except keep an alert eye out for embryonic blazes and pray that he could control them. The Chicago Fire Department numbered only two hundred men equipped with seventeen horse-drawn engines. This meager force was supposed to protect the entire city from fire. Williams had asked for more men and equipment an endless number of times, but the Chicago Common Council—which knew nothing of fire prevention—insisted that his department was adequately supplied.

Williams was even refused a fireboat for the Chicago River—a short-sighted restriction because the river flowing through the city was flanked on both sides by warehouses and wooden docks. There were twenty-four bridges crossing the river, and all made of wood. As if this were not enough, the Council made protection of the entire dock area virtually impossible by leasing the river street frontage to businesses, making the river inaccessible to fire engines.

In the first week of October, Chicago received warning enough that a major conflagration was possible: thirty fires broke out in various parts of the city, and Chicago's weary fire-fighters were hard pressed to keep them from spreading. The worst one started at about 10:00 P.M. on Saturday, October 7, when a planing mill on the west side of town caught fire. The Fire Department's Little Giant Company No. 6 received the first alarm. This company had been fighting blazes all week, and the firemen were almost too exhausted to respond—but they did. By the time they reached the scene, the mill was gone, a lumber yard was aflame, and a high wind was spreading the ravenous fire eastward over a four-block area.

When Chief Williams, fatigued and discouraged, arrived on the scene with several additional companies, he saw at once that his fire-fighters, numbering 185, could not on their own contain the flames, and he pressed into service hundreds of people who had gathered to watch the blaze. But the fire continued to gain ground.

One building that seemed fated to become a casualty was Dan Quirk's Saloon. But Quirk was a quick-thinker, not to mention a generous soul. He invited a group of men into his threatened saloon and passed out free whiskey. When he had the group in a salubrious mood, he made a self-serving suggestion.

"If you wet down my building and save it from the fire," he announced, "I'll give my entire stock of liquor away."

The men rallied to the cause. Although slightly under the influence, they managed to wet down the walls with hoses and buckets of water carried from the river. The building was saved as the fire raged around it, and Quirk magnanimously distributed bottles of liquor to everyone who had helped.

The Fire Department, completely sober, fought the stubborn blaze for seventeen hours before it was finally conquered. It was three o'clock on Sunday afternoon before the first companies to reach the fire returned. Sixty firemen were taken to the hospital suffering from burns or smoke inhalation. Two horse-drawn engines were sent to the shop for repairs. Fire damage was estimated at $750,000.

On the Saturday night that this preliminary fire raged at its most furious, a strange event took place at Farwell Hall in downtown Chicago. George Francis Train, an author and lecturer of considerable note, was speaking to a large audience. At the end of his speech he made a prophetic announcement.

"This is the last public address that will be delivered within these walls," he said in the sepulchral monotone of a preacher condemning his flock to everlasting hell. "A terrible calamity is impending over the city of Chicago. More I cannot say; more I dare not utter."

Whether Train had a premonition of evil things, as he implied, or had merely noted Chicago's vulnerability to fire and decided to gamble on a prediction, no one will ever know. But Chicago soon learned that, faker or seer, he was at least right.

It was Sunday, October 8, 1871. Katie O'Leary, who, with her husband, Patrick, lived at 137 De Koven Street on Chicago's West Side, was worried. In a barn behind the house were five cows, a calf, a horse, and a wagon. One of the cows was sick, and its loss would be a severe blow to the family's economy. So at 8:30 P.M. Katie decided she would go out to the barn to check on the animal's condition.

Picking up a kerosene lantern to light the way, Katie O'Leary trudged to the barn. After examining the sick animal, she decided that she ought to apply salt as a remedy and returned to the house to get some. Without giving the matter much thought, she placed the lantern on the barn floor. While she was gone, the restless cow kicked it over. The flame immediately set the straw on the floor afire, and the blaze crept up the dry wooden walls to the mow, which was filled with two tons of hay. In moments the entire barn was aflame.

Katie saw the burning barn as she emerged from the house with the salt. Her screams alerted her husband and several neighbors. All rushed to the barn with buckets of water—a pitifully inadequate amount of water for the rapidly spreading flames.

A neighbor, Dennis Sullivan, nicknamed Peg Leg for the obvious reason that he had an artificial leg, became a hero of sorts. Rushing into the flaming barn, he grabbed the frightened calf. However, as he tried to lead it from the barn, his peg leg caught in a crack in the floor. Thinking quickly, he unfastened the artificial limb and, hanging onto the bawling calf, managed to escape from the building.

This story of the fire's beginning is the one that has gained the most credence. It may or may not be entirely correct. Another version has Peg Leg Sullivan starting the fire when he crawled into the O'Leary barn to smoke and drink liquor. Still another story insists that small boys in the neighborhood had sneaked into the barn to smoke. There was even a man who claimed that elements formed in the soil throughout the midwest by a comet many thousands of years in the past had made the ground combustible—a theory apparently designed to intrigue Chicagoans of scientific bent.

Katie O'Leary herself confused the issue. She first admitted she had gone to the barn with the lantern and that the cow had knocked it over. But later, when she realized that damage suits might be levied against her, she said she had not left the house at all but had been in bed with her husband when the fire broke out.

Whatever the true story, one fact is unquestioned—the Great Chicago Fire *did* start in the O'Leary barn, and the idea that a contentious cow kicked over a lantern to start it had sympathetic appeal as a story. Thus the O'Learys went down in Chicago history and the unnamed bovine became the most famous cow of all time.

The Little Giant Company No. 6, which had fought the Saturday night blaze for seventeen hours, was the first unit of the Chicago Fire Department to learn of the new blaze. Those firemen who had not been injured in the previous fire had been asleep for some four hours only when the watchman scanning the city saw the roof blow off the O'Leary barn. He rang the alarm to awaken the men.

The firemen threw off slumber reluctantly, struggling into their uniforms and cursing their luck.

"Are we the only goddamn company to ever get a call?" one of them complained.

Men scrambled onto the fire engine as others harnessed the horses.

"Where is this one?" somebody asked.

"About five blocks from here," the watchman said. "Looks like De Koven Street."

As the Little Giant Company started for the O'Leary home, the fire began to make headway. A southwest wind caught sparks from burning hay and timber and hurled them toward other structures. In a few minutes a second barn, a paint shop, and a flimsy wooden home were afire. Flames leaped and danced in the gathering dusk.

It was at this point that things began to fall apart. Bruno Groll, owner of a drugstore about three blocks from the blaze, ran to a fire alarm box that had been installed by the Fire Department only a month before. As soon as he had pulled the alarm, Groll felt that he had done his duty. What he didn't know, however, was that the wiring of the system was defective. No message reached the Fire Department Headquarters in the Courthouse downtown. Only the Little Giant Company was aware of the blaze until the fire watcher in the Courthouse tower spotted the orange-red flames in the sky.

Even so, help could have been on the way quickly had it not been for an error on the part of the fire watcher. The flames from the new fire were shooting so high into the sky that he could see them plainly. But the watcher misjudged the distance to the fire. He estimated that the fire was at the corner of Halsted and Canalport, more than a mile from the actual blaze. He fed this information to the telegraph operator who alerted every fire company in the city.

The result of this unbelievably bad guess was that every fire company in Chicago, with the exception of the Little Giant, raced to a corner more than a mile from the scene of the fire—losing enough time, as it turned out, to doom the entire city.

Meantime, the Little Giant Company had reached the scene of the blaze and gone into action. William Musham, in charge of the company, found at once that his fire hose was of little help. The blaze had gotten a fast start and, whipped by the southwest wind, was moving rapidly north. He needed help from other companies, and he needed it quickly.

"Where are they?" he demanded of no one in particular.

The answer to his question was that they were all tangled up on the corner of Halsted and Canalport, and only now were becoming aware that the fire was a mile or more away from their position.

Someone else who had belatedly become aware of that fact was the fire watcher himself. He decided he had been wrong the first time and made a second guess.

"Correction!" he said to the telegraph operator. "I fix the fire at the corner of Twelfth and Johnson!"

This was closer, but still about seven blocks away. The corrected message might have helped the fire companies a little, but the telegraph operator refused to send it.

"It will only confuse them," he insisted. "They can see the location of the fire from where they are now."

This was true, but they had a long way to go to get there.

By the time the other companies arrived at the actual scene, the fire had gained a head start that was impossible to overcome. Two square blocks of houses and barns were already ablaze.

Adding to the hopelessness of the situation was the fact that one company's hose burst, another unit's hose burned up, and a third found pressure too low to pump water.

As the chaos mounted, frightened residents in the neighborhood of the fire fled from their homes or dragged their more precious belongings into the streets, where they impeded the movement of the fire apparatus. The southwest wind increased, sending sparks and flaming debris north toward the downtown business district.

Six blocks north of the O'Leary home was St. Paul's Catholic Church. Next to it was Bateham's Lumber Mill, on the edge of the Chicago River. Alongside the mill were two furniture factories. All of this was delectable fodder for the hungry flames approaching from the south.

Instead of burning its way north slowly, the windswept blaze hurled a burning timber six blocks through the air to strike the towering steeple of St. Paul's Church. Immediately the steeple caught fire, and in minutes flame engulfed the entire church. Bateham's Lumber Mill was next; a half-million board feet of lumber, one thousand cords of kindling wood, and almost a million wooden shingles went up in flames.

In the midst of this holocaust was Fire Marshall Robert Williams, directing his firefighters in what seemed a hopeless cause. He was a man sick at heart. Better than anyone else, he knew that all his predictions had come true, that the fiery

Armageddon he had foretold was upon them. This was the ultimate fire, and there was no way to keep it from destroying the entire city.

Within an hour after the original blaze had started in the O'Leary barn, there were three major fires in the city. The initial fire had split into two columns, and they were rushing north at a terrifying pace. The third fire was centered on the Bateham Lumber Mill.

It was now dark. The firemen struggled in the eerie, dancing light of the flames. People in the streets screamed and ran. The question was—where? The pursuing holocaust was everywhere. The draft caused by the flames, plus the southwest wind, became so strong that many people were blown down in the streets. Fiery sparks were everywhere, raining down on people as they fled. It seemed to many that the whole world was aflame.

As expected, the three fires finally merged, then whimsically started a fourth fire elsewhere. Near midnight the violent wind hurled several large pieces of burning wood a quarter of a mile across the river, depositing them on a newly completed stable belonging to the Parmalee Omnibus and Stage Company. There were no horses in the stables as yet, but the building had been stocked with lofts of fresh hay, which instantly caught fire. This new fire immediately spread to the nearby Chicago gas works, creating a new element of danger. Two huge gas storage tanks were directly in the conflagration's path. Tragedy was avoided only when a night superintendent at the gas works quickly pumped the gas into reserve tanks some distance from the approaching blaze, putting out every light in South Chicago.

But the savage fire was not to be denied. From the gas works the flames leaped to a nearby armory, and the cacophony increased as ammunition caught fire and exploded, adding to the terror of the night.

By midnight the uncontrolled fire was ravaging the heart

of the city. In his courthouse office, Mayor Roswell Mason realized that the Chicago Fire Department alone could never hope to extinguish the inferno. What was needed badly was outside help, and Mayor Mason began telegraphing other cities for aid. The response was immediate. Fire equipment was loaded on railroad cars in Aurora, Illinois, St. Louis, Cincinnati, and Milwaukee, and dispatched with all speed to Chicago. New York City was the most distant point from which help came—in the form of a special train carrying men and supplies.

While this help was on the way, two attempts were made to blow up buildings and create a firebreak to stop the spreading flames. Both failed. The fire simply raced through the rubble caused by the explosions and ignited buildings beyond. It seemed that nothing could stop it.

At one o'clock in the morning the courthouse itself fell prey to the conflagration. A blazing piece of timber struck the tower, setting the building afire. Flames quickly spread to the lower part of the building, where more than three hundred convicts were quartered. This caused a wrenching problem. Should dangerous criminals be turned loose on the already frightened populace? Should they be confined to their cells with flames raging around them?

The agonized cries of the prisoners could clearly be heard above the roar of the flames, and moments before the entire building collapsed, all prisoners—except murderers—were released to fend for themselves. The killers were handcuffed and led to the shore of Lake Michigan, where there was temporary safety. Those turned loose at once reacted in criminal fashion by looting stores, keeping just a step or two ahead of the advancing flames as they ran.

Shortly after the courthouse disaster, flaming debris hit a notorious section of Chicago known as Conley's Patch, the habitat of the city's gamblers, crooks, and prostitutes. The area's rickety wooden buildings burst into flames within seconds. Few escaped. Gamblers and bartenders and whores were turned into pillars of flame before they had time to flee their buildings.

Next the fire virtually consumed the business district. Two new plush hotels—the Bigelow and the Grand Pacific—were destroyed before they had even opened for business. The Palmer House, built by Potter Palmer and later to be rebuilt as one of the midwest's most distinguished hostelries, went up in flames. The famous Sherman House was also gutted. An older hotel—the Tremont House—succumbed to the fire, but its intrepid owner proved to be one of the more fortunate men to survive the destruction.

John B. Drake escaped from the Tremont House clutching a sack full of valuable silver and the money from his safe. The streets were filled with frantic people rushing around helplessly, each trying to find a place of safety, as Drake weaved his way among them on his way southward to his home. When he reached Congress Street, he noticed that the Michigan Avenue Hotel still stood. Unscathed as yet, it was right in the path of advancing flames and looked as if it were doomed.

Drake exhibited his Yankee ingenuity by walking into the lobby and confronting the distraught manager of the hotel. He introduced himself gravely, shook hands, and said: "I want to buy this hotel."

The manager's mouth fell open. This person was mad. In minutes now the hotel would be in flames. Guests and help were fleeing the scene.

"You can't be serious," the manager said.

"I'm deadly serious," said Drake, peeling off a thousand dollars and placing it on the desk. "That's a down payment, and I'll sign an agreement with you to buy the hotel if it survives the fire."

It was probably the fastest financial deal of its size ever to be consummated. The agreement was written up hastily and signed. Drake tipped his hat to the manager. "Thank you, sir," he said, and walked out of the lobby. He wended his way home without looking back. The Michigan Avenue Hotel would either survive or be destroyed. Either way, it was a worthwhile gamble.

As it turned out, the hotel was saved and Drake later completed the deal for the hotel's purchase.

Throughout the long night the sky over Chicago was alight with a saffron glow, and it was said that a man twenty miles away read a newspaper by the light of the flames. Frantic mobs roamed the streets, carrying belongings with which they hoped to renew their lives later. Looters broke display windows and stole jewelry and other easy-to-carry treasures. Liquor stores and bars were broken into. Brawling and fighting broke out among the criminal elements of Chicago's polyglot society. Some attacked innocent citizens, beating them to the ground and stealing their most precious belongings.

Many people did not escape the flames. One little girl's hair caught fire, and a drunk, trying to be helpful, tossed a glass of liquor on her to quench the flame. The alcohol blazed with a bluish flame and the girl was incinerated.

People living in second story apartments threw boxes and trunks out of the windows in frantic efforts to save their possessions. Many people were struck by these falling objects; some were killed. Draymen, trying to capitalize on the misery of the people, agreed to haul property to safety for $100. Few people had that much money in their possession, and those who did were badly treated. The draymen hauled the goods a few blocks, then demanded another payment to move it still farther away from the flames.

With the fire raging on both sides of the Chicago River, many people found themselves trapped on the wooden bridges. Caught in a pushing, angered mob, some fell into the river and drowned. Barges and other vessels, cut loose from their moorings and aflame from bow to stern, drifted in the river.

Many people tried to bury their favorite belongings in the ground. Some lugged their possessions to the edge of Lake Michigan where they dug huge holes in the sand and buried them. Others dug holes in their back yards to accommodate their treasures. A few actually buried pianos. One bride attempted to save her wedding presents this way, but they were fused by the heat of nearby flames. At least one man buried his wife in the sands of the Lake Michigan shore, leaving her room to breathe while he hauled water from the lake and poured it over the sandy mound to keep her cool.

The business district of Chicago was in chaos. The heat was so intense that stone buildings collapsed or exploded outward. Iron girders melted. Street car tracks curled into weird pretzel-like shapes. Streetcars burst into flame; the iron wheels melted into shapeless gobs of metal.

Dedication to duty was displayed by the newsmen of the city even as the flames closed in on them. Joe Medill, editor of the *Chicago Tribune,* tried desperately to get out an issue of the paper. The new *Tribune* building was thought to be fireproof, and in intense heat that made the inside of the building almost unbearable, reporters tried to write the first accounts of the monstrous fire. But their stories never saw print. The heat melted the press rollers, and a few minutes later the entire building burst into flame. The reporters fled into the streets to join thousands of others.

The final irony occurred when the new, elaborate waterworks building—in which Chicago not only took pride but had great confidence—was struck by a fiery piece of wood and set ablaze. The building and its pumps were destroyed within minutes, leaving Chicago without water except for what could be laboriously hauled from the river and the lake.

Monday morning, October 9, dawned over a desolate and still-burning city. The fire had reached Rush Street on the North Side, and factories and homes were burning there just as they had all night long in other areas of the beleaguered city.

More and more people had congregated along the lake, which seemed to be the only place of relative safety. Some, seeking an advantage, fled to the long North Pier that jutted into the lake, but flames cut them off and they had to be rescued by lake tugs. A suffocating heat settled over the beach area, and almost everyone spent most of the day in the water.

At the prestigious Chicago Club—a haven for wealthy businessmen—a note of strange courage emerged. The club had somehow escaped the fire, and a group of businessmen, many of whom had lost everything in the fire, gathered together at a champagne breakfast to toast their defiance. But the fire was vengeful. Before the men could finish their breakfast, the club was swept by flames. On their exodus from the building the businessmen picked up the plush couches in the lobby and as much champagne as they could carry and walked to the shores of Lake Michigan, where they sat in comical dignity and finished their meal.

It was not until Monday afternoon that the fire showed signs of burning itself out. Only in the northern sector of the city did the flames continue to spread; the rest of Chicago north of Mrs. O'Leary's barn had been reduced to smoldering, smoking rubble.

Then, on Monday night, a merciful but very tardy rain began to fall. Although it helped to quench the fire, it was ironic that the rainfall should be timed so badly. All summer long the city had prayed for rain. But it had not come until the city was destroyed. Had it arrived two days earlier the fire that started in the O'Leary barn might never have spread so rapidly—and Chicago may have been saved. Instead, it came only to quench flames that had already inflicted unparalleled destruction on the city.

The fire lasted twenty-seven hours. A total of 2,124 acres in the heart of the city was burned over. Eighteen thousand houses and buildings were destroyed, leaving ninety thousand persons homeless. Property damage was fixed at $196,000,000, approximately a third of the total wealth of the city. The exact number of dead will never be known. Most authorities, including the Chicago Historical Society, place the death total at 250. Others say 300. In any case, the death toll seems remarkably small considering the duration of the fire and the widespread destruction.

The next morning, Tuesday, October 10, trainloads of food, medicine, clothing, tools, and building materials began to arrive from other cities eager to help. Chicagoans, stunned by the terrible blow the city had received, doggedly went to work clearing the debris and rebuilding. City offices were opened in a church that had luckily escaped the flames. Five thousand "new" policemen were recruited and assigned to keep order in the city and to stop all looting. The army moved in with tents to shelter the homeless. Mayor Mason closed down all remaining saloons, forbidding the sale of whiskey or other spirits. He also fixed the price of bread.

Most remarkable of all, perhaps, the *Chicago Tribune*—despite the fact that its "fireproof" building had succumbed to the holocaust—got out a Fire Edition on Tuesday afternoon. Editor Joe Medill had located a small print shop with an old job press that had escaped the fire, and the story of Chicago's greatest tragedy reached a fire-weary public still anxious to read about what had happened to them.

In a front page editorial, the *Tribune* captured the feeling of Chicago citizens by stating: "In the midst of a calamity without parallel in the world's history, looking upon the ashes of thirty years' accumulation, the people of this once beautiful city have resolved that Chicago shall rise again!"

This spirit was amply exemplified by a real estate agent named William D. Kerfoot. He hammered together a wooden shanty from unburned scraps and opened his real estate office. On the front of the shack he placed a sign that read: ALL GONE EXCEPT WIFE, CHILDREN AND ENERGY.

It was the first building to rise from the ashes. In the next thirty days four thousand more would follow.

Chicago was on the way back.

2

Judgment Day in Peshtigo (1871)

THE famous holocaust that destroyed Chicago and killed 250 people in October 1871 was covered in long columns of type by every newspaper in the country. The entire nation was morbidly fascinated by the huge conflagration, and it was the topic of conversation on the streets and at dinner tables across the country.

It was not until several weeks later, however, that notice began to be paid to another fire, occurring some 250 miles to north of Chicago, one that devastated a much greater area than the Chicago fire and killed more than 1,100 people—on the very same day that Katie O'Leary's cow kicked over the lantern on the West Side of the Windy City.

Only one small newspaper covered the event in the thorough manner it deserved: the *Marinette and Peshtigo Eagle*. But that was because the tiny lumbering village of Peshtigo, Wisconsin, took the brunt of what has since become recognized as the worst forest fire in the history of the United States.

Big John Mulligan, a powerful man with bulging muscles, was foreman of a lumber gang in Peshtigo. On Monday, October 9—the day after the fire swept through Peshtigo—Mulligan stumbled into the village of Marinette, seven miles to the north. His clothing was scorched, his hands and face smeared with black streaks. He had a dazed, bewildered look in his eyes, and when people stopped him to ask what had happened, his voice came with sandpaper gruffness.

"Peshtigo is gone . . . burned . . . the whole town," he said. "Not a building is standing. People are dead in the streets."

An elderly woman inserted herself into the gathering group. "Our preacher told us yesterday that Judgment Day was coming," she said.

Mulligan nodded gloomily. "So did ours."

Judgment Day, as a matter of fact, had been heralding its arrival in northeastern Wisconsin all summer long. The great stretch of virgin timber northwest of Green Bay was dry from a six-month drought, and hundreds of tiny ground fires flourished in the dried-up marshes.

For most of the relentlessly hot summer the residents of Peshtigo were aware of the threat, but they found it difficult to believe that there was any serious danger. Marsh fires were commonplace and controllable. And besides, they thought, a couple of days of steady rain—which was sure to come soon—would eliminate any possibility of a major fire in the forests.

Peshtigo, built like a cozy bird's nest in the greenery of the forest a few miles from the shores of Green Bay, was a prosperous logging town in 1871. Most of its population of 2,000 was in the lumbering business, and all of its residents looked forward confidently to a bright future. There was solid reason for optimism. The Chicago and Northwestern Railroad was laying track that would connect Peshtigo with Milwaukee and Chicago to the south, and this would automatically insure greater wealth and growth for the town. Even now it was a thriving community, with 350 houses, three hotels, two churches, four saloons, a dozen stores, and a huge sawmill capable of producing 150,000 feet of lumber a day. One of its main industries was the Peshtigo Company, a large woodenware factory on the edge of the Peshtigo River. The river bisected Peshtigo, and the town had grown along both sides of it. A spiderweb of trails, cut through the surrounding woods, linked the city with other hamlets and with nearby farms.

An indication of potential tragedy came as early as April, but no one recognized its symptoms. Very little snow had fallen during the winter, April failed to bring its usual quota of rain, and the streams and rivers were low. May and June were also dry months. The sun blazed resolutely all summer long.

Lumbering interests were understandably worried about the dry weather. Profits, and the prosperity of the city, depended on the success of the summer-long log drives, and with the rivers low, there seemed no way to get logs to the mills.

Early in July a day of rain brightened hopes. If the rains would just continue now, all would be well. But the next day the scorching sun reappeared, and for two months not a drop of rain fell. Springs began to dry up; water became scarce. Not only was the lumbering business affected, but people had difficulty obtaining drinking water for themselves and their cattle.

During late summer the number of small fires in the forests increased. Deep in the woods, as if hiding until the proper time to reveal themselves, these fires smoldered in dry peat bogs and the tangled roots of the swamps. In August it was decided that an elementary precaution should be taken to make sure the fires did not approach too close to Peshtigo. A fire line completely circling the city was carved out of the forest. Only a tremendous blaze would be able to leap across the fire line, and this, naturally, was not expected, while rain, which would surely come in the fall, would put out the still-smoldering swamp fires.

On September 5 rain did fall. But it was only a teasing drizzle, and it simply turned the forests into a steaming jungle, doing nothing to alleviate the fire danger.

Meantime, two things happened to increase the danger—both of them man-made. First, in hopes of rain and therefore full waterways the lumbering industry was still cutting down trees, and useless limbs and slashings covered the floor of the forest. Second, railroad crews cutting a path through the forest south of the city were burning the debris—a dangerous practice in view of the dryness of the woods.

By mid-September even the most optimistic residents of Peshtigo were apprehensive. Small fires were becoming more frequent throughout the surrounding timberland. A blue-gray haze hung over the forest. The smell of smoke was in everyone's nostrils.

On September 24 a fire was discovered on the perimeter of Peshtigo, and every able-bodied man was called on to fight the blaze. Battling all night, the men finally extinguished the fire. The threat to the town was gone, but the memory of that night of fire-fighting lingered on. Everyone knew, now, that the surrounding forests could erupt at any minute. The earth was ankle-deep in slashings; logs had piled up in the shallow streams; dried-up marshes gave off dangerous gases.

Then came Sunday, October 8, 1871, a chilly day. People on their way to morning services at the churches noticed that a pall of brownish smoke lay over the city. The pitiless sun was unable to penetrate the haze.

Eyes smarting, the people poured into the city's two churches, hoping to find solace from their worries. But the ministers had no such comfort to give. Both predicted that Judgment Day was at hand.

"I have prophesized," thundered one preacher, "that the day would come when God would punish man's wanton destruction of the forests. That day is coming near!"

The men and women went solemnly home for Sunday dinner, the grim words ringing in their ears. All afternoon they scanned the skies, hoping for rain. But they saw only the sun, showing palely through the thick smoke. In mid-afternoon a shower of ashes, carried by a southwest wind, filtered down on Peshtigo. Men gathered in the hazy streets to talk the situation over.

At seven o'clock in the evening the people returned to the churches for evening service. The danger seemed closer now. The wind was causing the smoke to whirl and spiral overhead. Ashes continued to fall in a steady dry rain. In the two churches they heard again the same warning, and the prayers of most of the people that night asked that the city be spared.

When the churchgoers left the service they could hear a sullen roar coming from the south. The ominous sound was one that these people of the forest immediately recognized—the raging of flames not too far away. Frightened, they scurried to their houses and closed the doors and windows against the sound and the smoke. Somehow it seemed safer at home.

Meantime, people living on isolated farms scattered through the forest were in deep trouble. For several weeks they had lived in dread of what might happen. Not only were their lives endangered by the creeping fires, but they faced economic disaster if their farms were burned and their livestock killed.

James Langworth was an example. He had carved ninety acres and a comfortable living out of the woods around him. Now all that was threatened by a dry season and an approaching fire, though on Sunday, October 8, he felt no immediate need to leave his holdings to the mercy of the flaming forest. His wife and children were visiting friends in Canada, and he was alone. He decided to remain on his farm until flames drove him off, but he really didn't expect that to happen. He judged that the fire was at least a mile away, and he was doubtful that it was headed in his direction. Besides, there was a low swampy area between the fire and his farm, and Langworth was sure that if the flames reached the swamp it would act as a firebreak.

Still, Langworth noticed on Sunday morning that the smoke from the fire had grown thicker and that his stock, which consisted of a lone cow and several calves, was apparently suffering from smoke inhalation. He decided to free the animals and let them escape if they could, but he was stubbornly determined to stay, hoping to save his farm in some manner if the flames came closer.

At dusk Langworth sat in his home, staring out the window at the murky haze that lay like a tired cloud over the ground. Suddenly he noticed that at least a dozen rabbits had invaded the open area around his home. Squirrels, chipmunks, and raccoons followed. It was a forewarning that brought him up short. The forest creatures were fleeing from the advancing flames. That meant, almost surely, that the fire was heading toward the farm.

Langworth remained awake all night, listening with mounting dread to the macabre serenade of flames approaching his house, and in the early morning he decided he had no recourse but to leave. Tossing a few valuables into a blanket, he threw his belongings over his shoulder and started out. As he left the farm, his barn exploded in flames.

He had waited until the last minute, and now it would be a matter of outrunning the pursuing holocaust.

It was touch-and-go for a long time. Langworth stumbled awkwardly along a road that led to the nearby village of St. Charles. He could feel the heat of the merciless fire behind him as he ran. Flaming cinders fell upon him, and he constantly slapped at his clothing to extinguish sparks that threatened to set him afire. As he fled down the road, he noticed with dismay that the fire had eaten its way into the forest at his right, as well as behind him. Only the untouched forest on his left and the path ahead were still clear.

Then, all at once, tongues of flame leaped across the roadway, and he had no alternative except to turn left into the only remaining part of the forest that was not ablaze. He was almost exhausted when he suddenly emerged into an open area where a gang of workmen were clearing a fire line to protect the village of St. Charles.

Many other farmers were less fortunate. One man placed his wife and five children in a wagon and tried to outrace the fire, but the flames overtook them, killing everyone except the farmer himself, who managed to survive terrible burns. Another farmer and his family huddled in a clearing where they thought the fire would not reach them, but the heat from the surrounding flames incinerated them.

One distraught man, seeing there was no escape from the inferno approaching his farm, killed his family and then committed suicide. Another shot his livestock and then himself.

Meanwhile, the killer fire roared on, as if feeding greedily on these tragedies.

Back in Peshtigo, stark disaster on its way, most of the people in town huddled in their homes, uncertain and fearful. Fathers made a show of bravado, trying to assure their families that the fire would not strike the city. Mothers put their children to bed unconvinced that they would spend the entire night there.

At 9:00 P.M. the bells in the church began to toll mournfully. It was a signal that great danger was at hand, and the men of the village left their homes to do whatever had to be done to stem the advance of the red tide. An angry crimson glow brightened the sky to the south; fiery sparks descended on the village. Hastily, the men were organized into two groups. Half of them began beating out the small fires that were starting in the city. The others grabbed shovels and buckets of water and headed for the swamps around the town and the fire line that had been previously constructed.

Those in the forest worked furiously to beat out fires that kept springing up underfoot, but it was a losing battle. Before long, it became clear that it was futile to waste time extinguishing ground fires. Besides, the smoke was so dense now that they could hardly see what they were doing. Stinging eyes and difficult breathing made it almost impossible for them to work, and before ten o'clock the men turned and ran for town.

The women in town had already taken matters into their own hands, but there was little time left. Plucking their children from their beds and wrapping them in covers, they fled with their youngsters into the streets. The city was already beginning to burn. Towering pillars of fire were fingering the sky just south of Peshtigo. Sparks had already ignited the tops of trees within the town limits. Bushes and shrubbery were aflame. Instinctively the women knew the town was doomed, that it would be only minutes before its buildings would burst into flame.

The Peshtigo River seemed the only place where safety from the coming inferno might be found.

The men rushing back into town had the same idea, and they followed their women and children as they ran for the river. While this mad exodus was taking place, a great gale swept over the town, whirling about and scattering a rain of fire everywhere. Sheets of flame, burning pieces of wood, fiery embers fell over the stricken village. Great balls of burning grass, uprooted from the swamplands, dropped on the city. Violent explosions of marsh gas pelted Peshtigo with more flames.

Some of the fleeing people became a panic-stricken,

unthinking mob, and as a result a tragedy occurred that might have been avoided. Fear-crazed men, herding their wives and children to what they hoped would be a safe refuge, decided to cross the bridge over the Peshtigo River. In their eyes it seemed that the portion of town that lay on the other side of the bridge was an oasis that had not been hit by the fire.

The tragedy in the situation was the fact that those on the opposite side of the river were equally convinced that it would be safer on the other side. The result was that two frantic, pushing mobs ran headlong into each other at the middle of the bridge, forming a milling, fighting throng that guaranteed its own destruction. As they fought, the wooden bridge creaked and groaned. Suddenly the span swayed sickeningly and collapsed, cascading its human cargo into the river. Many drowned in the cool of the waters—perhaps a more merciful death than that of others in the heat of the fire.

Meanwhile, people waded into the river, reasoning that by submerging themselves in the water they could escape the twin fires that ate away the town on both sides of the stream. But as it turned out, some of these people also succumbed to the blaze.

A few brave men tried to man the fire engine, pulling it from one blazing building to the next, but the struggle to contain the fire was hopeless. At one point a man who had been in the railroad office rushed up to the firefighters with dismal news.

"This fire is bigger than we thought," he announced. "We just got a message on the telegraph from Green Bay. Chicago is burning!"

The men looked at him in awe. One raised his face to the heavens.

"Maybe the preacher was right. This is Judgment Day."

Abruptly, a cyclonic wind swept the town and put the finishing touches to Peshtigo. If there remained in anyone's mind the thought that the town might yet be saved, it was blown away by the fierce wind that suddenly spread flame throughout the city. It bowled over houses like tenpins. Roofs exploded from the tops of houses and went sailing through the super-heated air. Burning trees were uprooted and tossed against buildings. And behind the heavy wind came a wall of fire that engulfed the town within minutes.

The men deserted the fire engine and fled for the river.

Others did not think of, or avoided, the river. At least seventy-five people took refuge inside a huge boarding house, somehow believing that this large building would escape calamity. But the intense heat from the flames ignited the building, and all its occupants perished. The woodenware factory became a solid bonfire as the holocaust wreaked its destruction. One church burst into a fountain of fire, its steeple emerging from the top for a brief instant before the flames ate it away. Homes, hotels, saloons, stores were consumed.

The very air was so heated that a human being could burst into flames without being touched by a spark. This happened to many people as they ran toward the river. People were humans one second, a mass of flames a split-second later, and finally charred corpses.

The fire played capricious tricks on the fleeing populace. One husky young man racing through the streets toward the river with his injured wife in his arms, hardly able to see where he was going because of the dense, acrid smoke, ran into another person and went sprawling to the ground, his wife hurled to one side. Frantically, he picked her up again and renewed his run for the river. Only when he was submerged to his shoulders did he set the woman down on her feet. To his amazement he was looking into the face of a complete stranger. He had picked up the wrong woman after his fall in the street, and his wife was now a blackened corpse sprawled in the road.

Another young man had better luck. At the time of the

fire his wife was ill in bed with a fever. Rather than trust her ability to run to the river, he pushed and dragged the bed in which she lay all the way to the river as his five children ran ahead of him. With superhuman strength he dragged the bed into the water to a depth that covered all of his wife's body except her pillowed head. With the children huddled around him, he managed to survive the night and save his entire family.

Mostly the scene was one of commonplace tragedy. As a blonde girl with long hair rushed toward the river, all at once her hair and clothes caught fire. The flame then left her, but she fell to the street dead, her body uncharred. Her lungs had collapsed from the onslaught of the heat.

In another case a man realized he was too far from the river to make it safely and decided to take refuge in a horse watering trough. He was boiled alive. Children, separated from their parents in the mad crush of fleeing people, wandered in the streets until they were turned to ashes by the merciless heat.

A young woman, losing all reason, attempted to run into the raging forest. Before she could be stopped, she was turned into a pillar of fire. Terrified cattle ran loose in the street and trampled children underfoot. One man, unwilling to face death by cremation, killed his wife with a knife, similarly dispatched his children, and then cut his own throat. Another man calmly fastened a noose around his neck and hanged himself in his well. A group of seventy people huddled in a cornfield were incinerated. And along the edge of the Peshtigo River several pregnant women gave birth to children during the fire, as if attempting to renew a vanishing population.

Even many of those who reached the river failed to survive. Waves of heat swept over the river, suffocating people in seconds. Lungs collapsed under the attacks of heated air. The water was cold, but many people attempted to keep their heads under water as much as possible, coming up only for a gasp of air. If they happened to emerge at the moment a wave of heat crossed the river, they collapsed at once. Others, standing in the water, were struck by burning cinders or pieces of blazing wood hurled by the wind.

All night long the fire raged through Peshtigo, destroying everything in its path, killing people by the hundreds. One of the most graphic personal accounts of the fire came from a survivor named G. J. Teasdale. He said:

> During the day—Sabbath—the air was filled with smoke which grew dense toward evening, and the air grew warm, and hot puffs were quite frequent in the evening. About 8½ o'clock at night we could see there was a heavy fire to the southwest of the town, and a dull warning sound like that of a heavy wind came up from that quarter. At nine o'clock the wind was blowing very fresh, and by 9½ a perfect gale. The roar of the approaching tornado grew more terrible at 10.
>
> When the fire struck the town it seemed to swallow up and drown everything. The fire came on swifter than a racehorse and within twenty minutes of the time it struck the outskirts of the town, everything was in flames. About the time the fire reached Peshtigo House [a hotel], I ran out the east door and as I stopped on the platform the wind caught me and hurled me some distance on my head and shoulders, and then blew me on my face several times in going to the river. Then came a fierce, devouring, pitiless rain of fire and sand, so hot as to ignite everything it touched. I ran into the water, prostrated myself and put my face in the water and threw water over my back and head. The heat was so intense that I could keep my head out of the water but a few seconds at a time for the space of nearly an hour. Saw logs in the river caught fire and burned. A cow came to me and rubbed her neck against me and bawled piteously. I heard men, women and children crying for help, but I was utterly powerless to help anyone.
>
> Within three hours of the time the fire struck, Peshtigo was literally a sand desert, dotted over with smoking ruins. Not a hencoop or dry goods box was left. Through the sugarbush the case seemed to be even worse than in the town, as the chances of escape were much less than near the river. But great numbers were

drowned in the river too. Cattle and horses were burned in their stalls. The Peshtigo Company's barn burned with over fifty horses in the stable. A great many men, women and children were burned in the streets, and in places so far away from anything combustible that it would seem impossible they should burn. But they were burned to a crisp. Whole families, heads of families, children were burned, and remnants of families were running hither and thither, wildly calling and looking for their relatives after the fire.

For those who survived the heated air, the falling timbers, and the cold of the river, the flame-tinted night was a nightmare they would never forget. By dawn on Monday the fire had passed. There was nothing left of Peshtigo, but the flames still raged through six counties in northeastern Wisconsin.

Slowly, painfully, the survivors in the river crawled ashore. For a long time they sat numbly on the sand, not knowing what to do. There was virtually nothing left of forest or town. The forest had been reduced to black, smoldering tree trunks. Peshtigo had vanished—except for one lone cross in the graveyard and one wall of a burnt building.

After the roaring holocaust the silence was eerie. No sound of human habitation could be heard—no dogs barked, no cows mooed, no horses neighed. There was not even the cheery chirp of birds greeting the morning. Not even the fish in the river had survived; they rose dead to the surface and floated down stream. All was silence.

Finally the exhausted survivors dragged themselves to the streets of the now nonexistent city to search for loved ones. But few were successful. Most of the people caught in the fire were burned to a crisp; only charred, blackened bodies, unidentifiable, lay on the ground.

It was afternoon before help came to the ruined city. A steamer from nearby Marinette—which had miraculously escaped the general inferno—arrived in Peshtigo with food and clothing. It also brought dreadful news: twenty-three towns destroyed—Casco, De Pere, White Rock, Ahnepee, Elm Creek, Forestville, Little Sturgeon Bay, Lincoln, Brussels, Rosiere, many more—18 other towns severely damaged, and an as-yet uncalculated loss of human life.

Ironically, on Monday evening the much-prayed-for rain fell, for the first time in many months—heavy drops soaking deep into the smoldering earth. The beginning of 15 years of the most bountiful rainfall in the history of Wisconsin had arrived one day too late.

Exhausted by their ordeal, the people of Peshtigo nevertheless did not give up. That winter they cleared the debris from the streets of the town and the nearby forests. By spring they had planted crops in the earth that had once belonged to the woods. During the summer they cultivated their new lands, rebuilt their homes, churches, and public buildings. Within three years they had recovered completely—and were well on the way to establishing a reputation for dairy products that endures to this day.

Probably the most ironic twist of all was the fact that, because telegraph communication had been cut by the fire, the outside world for several days knew nothing about the calamity. When the news did filter through, papers gave it little space or failed to mention it at all.

The most thorough coverage was done by Luther B. Noyes, editor and publisher of the *Marinette and Peshtigo Eagle,* a newspaper that was only three months old. Noyes printed a Fire Extra on October 14 that told the terrible story of the Wisconsin forest fire and the fate of Peshtigo—and mentioned, in uncertain terms, that it was rumored that Chicago had some kind of fire too.

The Great Chicago Fire killed 250 people—and went down in history. America's most disastrous fire of all time, killing 1,182 city and forest dwellers, faded into obscurity. Judgment Day in Peshtigo came and went in deadly silence.

3

The Iroquois Theater's Fifteen Minutes of Hell (1903)

THE houselights of the richly appointed Iroquois Theater in Chicago dimmed for the second act of the Klaw and Erlanger musical extravaganza *Mr. Bluebeard.* In a cubicle dressing room backstage Eddie Foy, the famous comedian, struggled into the ludicrous costume of Sister Anne, a "poor but unhappy maid." On stage a double octette—eight men and eight women—moved gracefully into their opening song and dance routine as Joseph Dillea's pit orchestra struck up the first bars of "In the Pale Moonlight."

Lulled by the music and the dancing, few in the packed house noticed the sinister wisp of black oily smoke that coiled suddenly around the shell-like proscenium arch of the stage, curling upward toward the high ceiling of the auditorium. In the darkness of the house the ominous finger of smoke went unnoticed for several minutes—yet this was the beginning of what was to be one of the worst theater fires in American history.

The matinee audience of Wednesday, December 30, 1903, numbered more than two thousand persons, mostly women and children—1,724 of them seated, the rest occupying standing room—and it was not until burning pieces of scenery began to sprinkle the stage that the entranced audience realized that something was wrong. The dance routine wavered under the shower of fire, then abruptly fell apart. Two of the girls fainted, and the rest of the dancers broke ranks and fled into the wings. Eddie Foy, emerging from his dressing room at that moment, saw flame and smoke in the upper reaches of the stage and dashed instinctively to the footlights.

An uneasy murmur ran through the crowd. A few people rose to leave; others sat dazed and uncertain, watching with amazement the weird shower of sparks pelting the stage. Foy raised his hand to the crowd. "Stay seated," he said. "It is nothing. It will be out in a minute."

Then he looked down at Dillea, the orchestra leader. "Start an overture—start anything!" he pleaded. "For God's sake, play, play, and keep playing!"

Dillea whipped up the music. The audience teetered on the edge of panic, reason and fear fighting for control of their minds. For a moment it seemed that reason would prevail, but then somewhere a panicky woman cried: "Fire!" The dread word pushed the spectators over the brink. They rose in a body, and Foy, knowing now that nothing would stop them, pleaded with them to walk calmly to the exits.

By this time burning scenery was descending on Foy in a steady rain, and the stage was becoming an inferno. Retreating from the footlights, Foy shouted to the bewildered stagehands. "Bring down the asbestos curtain! Quick!"

Bill Sallers, a theater fireman, struggled to lower the curtain, but when the bottom was twenty feet from the stage, the right side caught on a reflector. The left side slid about twelve feet more, then stopped. The flames, as if angered by this obstacle to their progress, licked beneath it, reaching out hungrily for the fleeing audience.

It was altogether fitting that Benjamin H. Marshall should be proud of the Iroquois Theater—and of himself. At the age of twenty-nine, this clean-shaven, handsome young man had already established himself as one of Chicago's outstanding architects. He enjoyed looking back on his career, recalling how at nineteen he had been an office boy for an architectural firm and two years later had won himself a half interest in the business. Since then his career had grown and prospered, and there were those in architectural circles who were convinced he possessed the rare spark of genius.

The fabulous Iroquois Theater, with its French renaissance architecture, seemed to lend credence to this theory. Marshall had thrown all of his knowledge and talent into designing the glamorous theater, and he was satisfied that it was as near perfection as a building in 1903 could get. There was no doubt that it was the most beautiful theater in the Midwest. Precious few even in the east approached it, and Marshall knew that its plush appointments would be a major attraction to Chicago theater-goers almost regardless of what was playing there. But most important of all, it was fireproof—a little matter that in 1903 was, if not a rarity, at least an attraction to be highly advertised.

To Will J. Davis and Harry J. Powers, co-owners of the new theater and both veteran theater men, it had been obvious all along that the luxury of the theater would add to its drawing power. Davis, at fifty-nine, was recognized as the dean of Chicago theater. A thin man with sparse gray hair and a gray mustache, his dignified personage was well known in Chicago theatrical circles. The son of an Irish railroader, he had risen from a box office attendant in one theater to owner of several—evidence enough that he possessed a considerable talent for the business. Powers was a heavy-set man with dark hair pasted flatly over a rounded dome. Like many important men of the day, he sported a thick and curving handlebar mustache. He had begun his career as an usher, and now, at forty-four, he was recognized as an established theater operator.

Both men were peacock-proud of the new Iroquois, and if either of them had any doubts whatever about the theater, it was probably that the seating capacity of 1,724 might prove to be inadequate.

By November 10, when George Williams, building commissioner of the city of Chicago, a lean-faced man with fine straight lips that resembled stretched rubber bands, met with Marshall, Davis, and Powers to examine the building, he had already passed on the plans, especially the arrangement of exits, passageways, and aisles. Indeed inspecting the premises now was something of an anticlimax. Still, it was part of his job—and it *was* pleasant to associate with such prominent men as Marshall, Davis and Powers.

The four men made a complete tour of the building, inspecting the ornate foyer with its two magnificent staircases, the plush and tastefully decorated auditorium, the mammoth stage, and even the dressing rooms and the basement. When it was over, Williams indicated that in his opinion, the building would be completely fireproof when it was finished.

It is possible that Williams's favorable reaction caused Davis to forget—if, indeed, he had even remembered—the discordant opinion voiced three months before by a dedicated young man named William Clendenin, editor of a magazine called *Fireproof*. Clendenin had asked for permission to inspect the half-finished building. When his tour was over, he reported to Davis that there were several serious flaws in the building's construction. There was no sprinkler system over the stage and no ventilating flue over the stage to carry flames up and away from the audience. The skylight above the stage was nailed shut, heavy use was made of wood trim, there was no direct fire alarm connection with the fire department, and several other things were lacking.

As far as Davis was concerned, Clendenin was wasting his time. Blueprints of the Iroquois Theater had already been approved by Building Commissioner George Williams. The theater was to be inspected again before opening.

Clendenin walked out in a turmoil of anger and outrage. Back in his office, he wrote a scathing editorial in *Fireproof* magazine condemning the theater as one of the worst firetraps in the city. The article had no effect, however, except to provide a release for Clendenin's anger, for few people ever read *Fireproof*.

Apparently Davis attached little importance to Clendenin's warning or the editorial. Certainly, nothing was done to correct the oversights he pointed out. Indeed, Davis felt his own judgment in ignoring the upstart Clendenin was vindicated when Williams inspected the premises and gave it his approval "from dressing rooms to capstone." He was doubly sure when, the day before the Iroquois opened to the public, Inspector Ed Laughlin, of Williams's office, put his unqualified approval on the structure and declared it "fireproof beyond all doubt." The theater had more exits than any in the country—thirty in all, twenty-seven of them double-door fire exits; each floor—orchestra, balcony, and gallery—was equipped with emergency exits feeding into the foyer.

Both Williams and Laughlin ignored such matters as the absence of the stage sprinkler system, the fact that the skylight over the stage was still nailed shut, and many other violations of the building code.

Opening night, November 23, was a memorable event for Davis, Powers, and Marshall. All three had to answer the curtain calls of an appreciative audience. This so warmed the heart of Davis that when, two weeks later, a stagehand named Joseph Daugherty sounded another warning, it fell on reluctant ears.

A big man with gnarled and irregular features, Daugherty came into Davis's office almost apologetically to report that there had been a little fire backstage.

"A *what?*"

Daugherty hastened to explain that it was out and that some trash had simply caught fire. But he added that when the asbestos curtain was lowered, it had stuck on a reflector about twenty feet above the stage.

Davis asked if Daugherty had reported this matter to Bill Carlton, stage manager of the Iroquois.

Carlton wasn't around, Daugherty said. That's why he had come to Davis. In his opinion the position of the reflector should be changed if the curtain was to be lowered, even though the actors wanted the reflector in that exact spot.

Davis said he'd look into it, but when Daugherty disappeared through the door, Davis dismissed the incident from his mind. It was hardly an insoluble problem. Carlton would see to it.

December 30, 1903 was a gray, cold day in Chicago. People fortunate enough to have tickets to *Mr. Bluebeard* were glad to get inside and settle into the comfortable seats. The first act warmed them even more, and as the second act got underway, no one in the contented audience could have possibly foreseen the horror about to descend on them.

At approximately the time the second act of *Mr. Bluebeard* started at the Iroquois, architect Benjamin Marshall was at work on a new theater in Philadelphia. He had noted with delight that the Iroquois was drawing capacity crowds, even during the traditionally slow week between Christmas and New Year, and he was certain that the majesty of the theater itself contributed greatly to its drawing power.

His connection with the famous Iroquois had, of course, done him no harm. His reputation had skyrocketed with that of the theater, and his services now were even more in demand.

On the same fatal afternoon, Commissioner George Williams sat in his office, looking out at the blustery wind sweeping through Chicago's dreary streets. He was feeling particularly contented this afternoon. He had held the job of city building commissioner for two years now, and it was an easy and thoroughly comfortable job. His predecessor had quite literally quit under fire—following the tragic burning of St. Luke's Sanitarium in which twenty mental patients, strapped to their beds, perished. Williams had been selected as the proper man to see that no such fire ever again occurred.

Well, he had lived up to his billing. There had been nothing to criticize in his two years on the job, because he had surrounded himself with capable and reliable inspectors—perhaps not as many as he would have liked, because department funds were low, but as many as he could get. If he thought at all of the Iroquois Theater that afternoon, it must have been with complete satisfaction. He was convinced that if everybody who erected a building in Chicago built one as near perfection as the Iroquois Theater, his job would hold practically no perils.

At almost the same moment Davis, who that afternoon had attended a funeral of a friend, set out on his return journey to his offices at the theater. Powers had remained in charge, and he stood in the foyer contemplating the crowd with quiet contentment. Just then Bill Curran, one of Williams's "capable and reliable" inspectors, walked through the foyer.

"Everything all right, Inspector?" Powers asked, thinking that Curran might object to the standing-room crowd.

"You've got a fine house," Curran said. "The aisles are clear and you're handling the crowd in good shape."

This was five minutes before the holocaust.

Backstage, at this moment, several other things were occurring that would compound the tragedy. Bill Carlton, stage manager, left his post behind the curtain and walked to the foyer, where Powers noticed him.

"Aren't you supposed to be backstage?" Powers asked.

Carlton grinned sheepishly. "I thought I'd like to view the show from up front once."

"Don't blame you—damned good show," Powers said goodnaturedly.

Backstage several stagehands stood near the door. Having finished setting the stage for the second act and having nothing to do until the end of the performance, they slipped out of the door and headed for a nearby saloon.

At about the same time Edward Cummings, a carpenter who was in charge of the electrical mechanism controlling the asbestos curtain, remembered that he wanted to make a small purchase at the hardware store down the street. Having seen the stagehands leave for what he considered a less worthy reason, and satisfying himself that everything backstage was in order, he left the building.

In contrast, one man was frankly worried. He was William McMullen, a stagehand in charge of one of the calcium spotlights in the wings. Above his light a "tormentor," or bit of scenery representing foliage, swayed back and forth, occasionally scraping the hot metal surface of the light. As the second act got underway, McMullen watched the flimsy piece of scenery uneasily.

Suddenly the spotlight sputtered. As McMullen stooped to attend to it, the tormentor caught fire. Frantically, McMullen tried to crush the tiny flame in his hand, but it writhed two inches out of his reach. As the flame twisted its way upward, theater fireman Sallers grabbed an extinguisher. He aimed it at the still-small blaze, but the stream fell inches short. Quickly other pieces of scenery caught fire—and there was plenty of it. Except for the acting area, the stage was filled with 40,000 cubic feet of scenery, wooden sticks, frames, paint and canvas; and 180 drop scenes hung with 75,000 feet of new, oily and highly inflammable manila rope.

Orange tongues of flame already danced in the upper reaches of the stage when Eddie Foy burst from his dressing room and raced to the footlights to implore the audience to remain calm. But the sight of the dancers stampeding into the wings and the cry of "Fire!" threw the crowd into panic.

At Foy's command fireman Sallers, unfamiliar with the asbestos curtain mechanism, attempted to lower it, only to have it catch on the reflector Joseph Daugherty had complained about to Davis several weeks earlier. Showered by blazing pieces of scenery, Foy at last retreated toward the stage door—and just in time. With a roar the stage loft collapsed, and a Niagara of fire poured down. In the thundering cascade the central switchboard was destroyed and the house lights went out, plunging the theater into a fire-flecked darkness.

Meanwhile the terrified dancing girls had reached the stage door, determined to quit the theater despite the cold weather and their flimsy costumes. A helpful stagehand threw open the door and the girls stumbled out into the cold. This action sealed the doom of the audience. The open door created a draft that bellied the half-lowered curtain and sent the flames roaring into the panic-stricken crowd.

Driven by the draft, the flames rolled high over the heads of those on the first floor, spanned the fifty-foot space to the balcony in one prodigious leap, and there split in two. The lower part swirled under the balcony and out into the foyer; the other half roared up to engulf the people in both balcony and gallery. In its destructive path through the darkened theater, the fire ignited everything burnable—draperies, curtains, wood trim, seats, and all decorations.

By this time the audience was engaged in a titanic struggle. In the eerie half-light of the flames, they fought and pushed their way toward the exits, clawing each other like enraged animals. In the first few minutes dozens of people were stripped naked by clutching hands. Many—some of them human torches—fell or were pushed from the balcony to the orchestra, smashing themselves, setting fire to those below. In the awful crush, those who fell were trampled on—and in the balcony and gallery, particularly, the frantic crowd trying to escape the horrible flames stumbled over bodies that quickly piled up four and five deep.

Some showed ingenuity and calm—and escaped. Winnie Gallagher, an eleven-year-old girl, avoided the shoving throng in the aisles by walking over the backs of the seats to safety. Others, showing equal imagination, were less lucky. Three young girls about ten years old ran to a corner of the balcony near the rail, thinking they were safe there—but a pillar of fire swept them over the rail to their deaths.

For the most part people exhibited only the instincts of fear-crazed animals. Vicious fights broke out in the mad attempt to reach the exits. Children were shoved aside by adults; and women were beaten physically by fear-maddened men. And at the height of this carnage several things happened to add to the tragedy.

On the first floor, sixteen-year-old ushers, who had received no instructions on handling an emergency, made up their own rules. Some actually held the doors against the surging crowd. Others, terrified, fled without opening emergency doors. The crowd swept against the doors, only to find that in several cases the doors were locked. One pair of doors opened inward, the other, outward—creating an insoluble problem for the excited mob. In those cases where the crowd managed to get the doors opened, they found that first floor door sills were four feet above the pavement. In the mass exodus people were pushed out, fell, broke limbs, and provided a human carpet for others to walk over.

Most of the eight hundred people occupying the first floor of the theater escaped because the fire rolled over their heads.

But in the balcony and gallery over half of more than 1,100 spectators died. Here the flames broke full upon the crowd as they frantically sought to locate exits in the blackness.

No exit signs helped them. Many emergency doors were unmarked; some were obscured by heavy draperies. As a result, each person tried to make his way to the entrance he had used before. Frightful congestion occurred in several spots while unmarked exits were overlooked. Those who managed to open emergency doors found that some fire escapes were not equipped with ladders to the ground. Many frightened people jumped from the ladderless escapes, only to break legs and arms in their falls and to provide a mattress for others to leap upon.

Opposite the western exit painters were working on a building across the alley. Seeing the plight of the theater throng, they placed planks across the alley from the theater fire escape to the opposite building.

"Walk across!" they shouted. "Walk straight across and don't look down!"

The people hesitated, reluctant to risk a fall, yet feeling the searing heat of the flames behind them. Two sisters, Hortense and Irene Lang, sixteen and eleven, finally tried it—and made it safely. In all, twelve people eventually escaped over the planks, but the last one was pursued by a pillar of fire that destroyed those still standing on the fire escape.

Will Davis returned from the funeral to find the street in front of the Iroquois clogged with the injured and the dead. Nearby restaurants and saloons served as makeshift morgues and hospitals. A priest was performing the last rites, and doctors, called hastily to the scene, worked swiftly to aid the injured. One man, his face contorted in agony, pulled at Davis's sleeve. "I have twelve children in two boxes and they're missing!" he cried. "Are they in there?"

Davis shook his head hopelessly. "I don't know," he muttered. Then he turned to Powers who stood shocked and bewildered in the street. "My God! How do you answer such questions?" Powers simply shook his head, too numb to speak.

In the sixty-foot-high gilt-and-marble foyer the final tragedy was being enacted. The main floor spectators had surged into the foyer like a great tidal wave. Those in the gallery who managed to escape the horror of the flames descended to the balcony and collided head-on with another mass of humanity trying to reach the foyer. In a vicious struggle, men and women were hurled down the stairs or were shoved over the iron balustrades to fall on the surging crowd below. Others pushed their way down the two ornate staircases, meeting the main floor spectators in the foyer to form a giant whirling vortex of humanity fighting its way to the main exit. Suddenly, the flames swept over the almost immovable mass.

When firemen later reached the scene, they found bodies so hopelessly entangled that it was difficult for them to separate even the dying from the dead. In the balcony, bodies were piled six deep in the aisles. Many of the people were nude, their clothing torn to shreds, others were horribly mangled and mutilated.

One man was so badly trampled underfoot that he had no skin left above his waist and his skull was bare. A woman who had fallen from the balcony to the orchestra was found almost severed in two by the back of a seat. Another man was bent backward over a seat, his spine broken. A husband and wife were found locked so tightly in one another's arms that the bodies had to be taken out together. The charred and blackened bodies of a woman and her child were found fused by the flames. An infant was discovered in a corner of the balcony, all clothing torn off except its shoes; another child was found decapitated. Some persons were still alive, buried beneath others whose bodies had taken the brunt of the stampede and flames.

One of the most tragic theater fires in history had lasted

just fifteen agonizing minutes. When it was all over, though the "fireproof" building still stood, its interior resembled a burned-out volcano. The stage, boxes, main floor, balcony, gallery, all wood trim, draperies and even the asbestos curtain—which was found to be made of little more than paper—were destroyed. A total of 591 people had perished, and hundreds more were badly burned or injured in the terrible crush. The property damage was estimated at the then brisk sum of $20,000.

When Benjamin Marshall received the news in Philadelphia he was dumbfounded. Later he said contritely, "I am completely upset by this tragedy, more so because I have built many theaters and have studied every playhouse disaster in history to avoid errors. But I suppose no building is fireproof as long as a stick of wood is used. I will never use wood in a theater again."

As a result of the holocaust, twelve people were indicted on manslaughter charges. These included owners Davis and Powers, Building Commissioner Williams, and various employees of the theater who were suspected of being derelict in their duties. Aroused by the horror, Coroner John Traeger began an inquest within a week. Although not indicted, Marshall testified at the hearings. Even the Honorable Carter Harrison, Mayor of Chicago, was dragged into the investigation when it was learned that he had given a report to the City Council that all theaters in Chicago were violating the fire ordinances—and then had failed to see that action was taken. No one, Traeger vowed, would be spared in an attempt to fix the blame.

The inquest—at which more than two hundred people testified over a period of three weeks—was a national sensation. It exposed unbelievable laxity on the part of theater and city officials charged with the public's safety—all of whom exhibited a remarkable agility at sidestepping responsibility.

Marshall insisted that, to his knowledge, the fire ordinances had all been complied with but admitted that the house was put up hastily and "perhaps some things were left undone that should have been done." Davis attempted to cover up his lax supervision of employees by shifting the blame to his subordinates. These men insisted that they had been given few instructions and that their duties were not clear-cut. Powers took refuge in the fact that he served only in an advisory capacity as a "resident owner and associate manager" and was not personally responsible for details of operation. Building Commissioner Williams tried to explain away his shortcomings by saying his department was under-budgeted and understaffed and could not perform adequately.

As the questioning continued, it became evident that the Iroquois fire was caused by a combination of three deadly factors—flaws in the building's construction, mismanagement of the theater, and apathy in the building commissioner's office.

On the theory that the owner was responsible for providing a safe place for an audience, Davis, rather than Marshall, was primarily blamed for the flaws in the building. These included the defects editor Clendenin had pointed out to Davis three months earlier—lack of a backstage sprinkler system and overhead flue, and the fact that the skylight, which might have served as a flue, was fastened shut. In addition, there were the dangerous first floor door sills with their four-foot drop to the ground, the ladderless fire escapes, and the fact that neither the balcony nor the gallery had independent exits to the street, resulting in the tragic crush in the foyer.

Owners Davis and Powers were proved to be completely incompetent in their management of the theater. They had ignored advice from Clendenin, ignored Daugherty's complaint about the asbestos curtain's sticking, and failed to instruct the under-age ushers in their duties in an emergency. Contrary to fire regulations, the stage area was loaded with burnable material, the theater was overcrowded, several emergency exits were locked and others hidden by draperies, there were not enough suitable fire extinguishers on stage, no fire hose attached to standby pipes, and no direct fire alarm connection with the fire department. In addition, they had failed to control their employees since several key people—including stage manager Carlton, carpenter Cummings, and many stagehands—were not at their posts when the fire started.

Building Commissioner Williams proved his incompetence by his own amazing testimony. He admitted that he had passed on the theater blueprints "after looking at them for about ten minutes." Subsequent reports received by him on the progress of the theater were in the form of brief memoranda noted by an inspector in a book Williams never looked at. The final report made the day before the Iroquois opened was an oral "okay" by Inspector Ed Laughlin.

"And you were content to accept a verbal okay on the Iroquois?" he was asked.

"It was the best I could get," Williams answered.

"Did you ask what was meant by okay?"

"I did not."

"Did you ask if all fire apparatus was in?"

"I did not."

"Did you ask about separate exits from the galleries?"

"No."

"Did you ask if the stage flue was in working order?"

"I did not."

"Did you ask about the fire alarm connection with the Fire Department?"

"I asked him *no* questions."

Later, when asked about his personal visit to the theater on November 10, Williams admitted he did not notice any violations of the fire ordinances. When pressed, he said that he had not read the ordinances until after the fire. Red-faced and angry, he insisted he was being "soaked for the faults of others" and that his inspectors had grossly misinformed him.

"Isn't it your duty to see that your inspectors are doing their job?" Traeger asked.

Williams shifted uneasily in his chair. "Once in awhile I have done so," he said weakly.

In the end, seven men, including Davis, Williams, the Mayor, and several theater employees, were held over for Grand Jury investigation. Davis was finally charged with failure to instruct employees and to see that the theater was properly equipped as required by the fire ordinances. Although, seemingly, Marshall should have shared the blame for this, he was not held on any count. Powers was excused because of his "advisory" status with the theater. Williams was charged with gross neglect of duty, and Mayor Harrison was deemed responsible for "lack of force" in dealing with Chicago's firetraps.

The Grand Jury, reviewing the evidence, subsequently exonerated Mayor Harrison and several others, but it upheld the manslaughter charge against Davis and two minor theater officials and charged Williams with misfeasance.

Despite all this commotion, no one ever served a jail sentence. After much legal finagling, Davis and others managed to get the charges quashed on the grounds of technical faults in the Chicago building ordinances. None of the victims or relatives of the dead ever collected a cent of damage. The only person to serve a jail sentence was a nearby saloon owner whose property had been used as a temporary morgue. He was convicted of robbing the dead.

The Iroquois tragedy affected Davis deeply, however, and he never fully recovered from the shock. He made a brave effort to continue in the business to which he had devoted his life, even attempting to renovate and reopen the Iroquois. But Chicagoans would have none of it, and the theater was finally sold and dismantled. Finally, at the age of seventy-five, in 1919

Davis fell ill and died. Some of the most prominent people in Chicago attended the funeral, and newspapers lauded him as one of the city's most illustrious theater men.

Powers continued in theatrical work until his retirement in 1930. He died in 1941 at the ripe age of eighty-two and was similarly praised by the press. Commissioner Williams, after his brief moment in the limelight, faded from public view.

Benjamin Marshall, despite the bad publicity arising from the Iroquois fire, went on to even greater success in the field of architecture. In Chicago he built such edifices as the Blackstone, Drake, and Edgewater Beach hotels, the Blackstone Theater, and various office buildings. Outside Chicago he designed Maxine Elliott's Theater in New York, the Forrest Theater in Philadelphia, the Edgewater Gulf Hotel in Biloxi, Mississippi, and many other places of note. By the time of his death in 1944 at age seventy, he was considered one of the country's most outstanding architects.

Despite the fact that those responsible for the Iroquois disaster escaped punishment, some long-range benefits to the public did accrue. Officials in other cities, shocked by the tragedy, began to inspect their theaters more closely. Existing fire ordinances all across the nation were more rigidly enforced, and new ones written to make America's theaters safer.

Today's theaters are much safer places. Current fire ordinances, which cover every conceivable situation, are the result of lessons learned from studies of all playhouse holocausts of the past—including fifteen minutes of hell in Chicago.

4

The Day the *General Slocum* Burned (1904)

CAPTAIN William H. Van Schaick stood at the rail of the big paddle-wheel excursion boat *General Slocum* and watched the passengers boarding from New York City's East River Pier at Third Street. He was not particularly excited about the day that lay ahead of him. Van Schaick, sixty-three years old, had been a licensed pilot for forty years and a ship's master for thirty, and today's excursion was just another in an endless chain of trips that had already taken up more of his life than he cared to recognize. In fact, this particular outing gave promise of being one of the most hectic and noisiest of the entire summer season because of the great number of children and mothers aboard. Loud children and chattering mothers were not numbered among Captain Van Schaick's favorite things.

It was Wednesday, June 15, 1904—a warm, sun-splashed day, just right for an excursion. This was the day chosen by St. Mark's German Lutheran Church for their annual family outing—a trip on the *General Slocum* up the East River into the Sound, where they would disembark at Locust Grove, Long Island, for a picnic, returning later in the day. When all were aboard, the group would number 1,358 people: some 750 babies and young children, their mothers, some fathers lucky enough to have the day off from work, a scattering of aunts, and a bevy of Sunday school teachers.

The high spirit of the occasion was revealed in the way the happy throng approached the ship. The Reverend George C. F. Haas, pastor of St. Mark's, had led them to the pier in a gay march from the church in "Little Germany" on New York's East Side. With them was a German band oompahing its way through the streets with ear-shattering gusto. It was 8:30 A.M. when the crowd started to board the *General Slocum*, and 9:30 when the boarding was completed.

By this time Captain Van Schaick had escaped from the laughing and singing crowd to the seclusion of the pilothouse. The gala throng had dispersed throughout the ship, thus assuring a general pandemonium from bow to stern. Children romped about, exploring every passage and companionway. A group sang German hymns under the direction of the smiling Reverend Haas. It was a festive occasion, held on a beautiful day by people who had shed their everyday cares and had nothing on their minds except having a good time.

There was, of course, no thought of impending disaster. The *General Slocum* was considered the finest excursion boat serving the New York area. A huge, white side-wheeler, two hundred fifty feet long and three decks high, she was capable of handling twenty-five hundred passengers. Owned by the Knickerbocker Steamboat Company, the *General Slocum* was thirteen years old (still young for a ship) and had recently been refurbished and repainted.

At a signal from Captain Van Schaick the lines were cast off and the ship's whistle sounded loudly, momentarily drowning out the singing of hymns on the hurricane (upper) deck. Moving out slowly as the sidewheels churned the water into foam, the *General Slocum* headed for the middle of the East River. The sidewheels revolved faster as the excursion boat picked up speed, and within minutes the vessel, now traveling at a steady 10 knots, passed Blackwell's Island and pointed its prow toward the narrow stretch of the river known as Hell Gate, just opposite 92nd Street.

The first evidence of a possible fire—in the hold of the vessel—was noticed by a young boy who saw, on the port side, a thin spiral of smoke drift up to the main deck from below. Beneath the main deck at that point was a storage room known as "the second cabin in the hold." Startled, the boy ran at once to the pilot house to alert the captain.

Captain Van Schaick was annoyed by the intrusion and told the boy, in no uncertain terms, to beat it.

A few minutes later Frank Perditsky, a fourteen-year-old youngster, also saw the smoke. He, too, ran to the pilot house, and received like treatment.

Actually, down in the storage room tiny flames with ambitions to become larger were licking away at an assortment of highly inflammable material. The forward cabin in question was used as a catch-all for a variety of odds and ends. Among the material stored there were a dynamo, two barrels of oil, a barrel of mineral sperm, paddle buckets, paint pots, camp stools, life preservers, old hose, and lamps. Just before departure of the *General Slocum* from its East River pier, two barrels of salt hay, in which drinking glasses had been packed, had been tossed haphazardly into the room.

Still moving at ten knots, the *General Slocum* plowed its way up-river for another half mile. On her port side she passed a dredge removing earth from the bottom of the river along the shore. A deckhand on the dredge noticed that smoke was coming from the *General Slocum*'s hold. Immediately he blew four short warning blasts of his whistle. This signal was picked up by skippers of other craft along the shore, all eager to apprise the excursion boat of its danger. Several of the craft then fell in behind the *General Slocum,* following it up the river and emitting four-blast signals constantly.

Captain Van Schaick, who had already committed one blunder by ignoring the warnings from the two boys, now perpetrated another. He blandly ignored the warnings from the other boats! Under the guiding hand of her mistaken captain the *General Slocum* continued its journey as if no danger existed.

Indeed, it was not until a third young boy spied the smoke that any action was taken, and even then the action was so amateurish that it aided, rather than diminished, the fire.

The boy, seeing smoke this time tinged with red flame, raced for the entrance to the bar on the main deck, seeking help. There he found John Coakley, a young deckhand who had been aboard the *General Slocum* only a week, huddled over a beer.

"The ship's on fire!" the boy cried excitedly.

Coakley, at least, was stirred to action. When the boy pointed out the location of the suspected fire, Coakley realized that it was probably centered in the storage room. He raced down a companionway and opened the storeroom door. The inside of the room was black with smoke, but through it he could see flames hungrily lapping at the hay in which the drinking glasses had been packed.

Coakley, untrained in the handling of fire aboard ship, spent several minutes trying to put out the fire. First he seized a canvas and tried to smother the burning hay. When this didn't work, he grabbed several bags and tossed them on the fire. But the bags were filled with easily ignited charcoal. Panic-stricken, Coakley then left the fire and rushed on deck, searching for First Mate Ed Flanagan.

It took him several more minutes to locate Flanagan, who immediately ordered Coakley to unreel the hose on the main deck. Then he ran to the engine room and told the engineer to pump water in the stand-pipe. Last, he called the pilot house on the speaking tube.

"She's afire," he reported to Pilot Van Wart. Van Wart at once relayed the message to Captain Van Schaick.

Van Schaick's face turned ashen. His hands trembled. His first reaction was to reach for the speaking tube and order the engine room to bring the *General Slocum* to a halt. Then he changed his mind, deciding to keep her on course until he had investigated.

When Van Schaick reached the main deck stairs to the storage room he found them blocked by flames. Quickly he returned to the pilot house. Confused and excited, he compounded the danger by making a third mistake, deciding to order full steam ahead with the idea of beaching the *General Slocum* on North Brother Island.

Van Wart looked at the captain with a startled expression on his face. North Brother Island was almost three miles ahead. At full speed—ten knots an hour—the wind off the bow would sweep the flames toward the stern and engulf the entire ship. Besides, there were many other places along the East River where a disabled ship could be beached.

Van Wart protested, but Van Schaick was adamant, saying that North Brother Island was the only safe place to beach her.

Finally, the captain sounded the ship's fire alarm. The clang of the gong was the first signal to most of the passengers that there was fire aboard. But before they could react to the alarm, a second, and very frightening, event occurred and pushed them over the brink into panic.

A woman on the main deck stepped over a floor grating and was immediately set afire by a flame that darted from the grating and ignited her dress. She ran down the deck, shouting "Fire! Fire!" at the top of her lungs until the flames consumed her.

That did it. Women, suddenly terrified, began racing

around the deck looking for their children, and in that first awful moment dozens of people fell and were trampled.

Meantime, Flanagan and his deckhands had managed to reel out the fire hose. It stretched like a dead snake along the main deck, its nozzle pointed in the direction of the stairway leading to the storage room. Flanagan ordered full pressure, the water stiffened the hose but never reached the nozzle. Instead the hose split along almost its entire length, like a sausage cut with a knife.

In the pilot house Captain Van Schaick stood at rigid attention like a condemned soldier awaiting a volley of shots from a firing squad. The *General Slocum* was approaching the mouth of Hell Gate. A brisk wind came off her bow, fanning the now-spreading flames. His crew was in a state of shock and disorganization. The captain did nothing, even failing to sound an alarm for help to either shore. Stubbornly he kept the *General Slocum* plowing upriver at ten knots an hour toward North Brother Island, still almost three miles away.

As the excursion boat steamed into Hell Gate, the wind off the bow began to sweep the flames toward the stern. Huge clouds of black oily smoke, interspersed with dancing flames, swept across the main and lower decks. The people fled toward the stern, the only part of the ship that still gave them refuge. In the crush of humanity, families were separated and the anguished cries of children and their mothers, each searching for the other, could be heard across the decks. Some people were killed in the rush to the stern; some fell overboard; others, slower, were overtaken by the flames.

The *General Slocum* had become a death ship.

There was little help from the crew. A few deckhands tried to free the life preservers, which were wired to the deck rails and stored in racks overhead, and the passengers fought each other to get them. But most of the preservers were useless, disintegrating in the hands of those who tried to put them on.

The straps and canvas tore, and the cork inside the preservers, pulverized by age, filtered through the people's hands and sprinkled the deck. An attempt was made to lower the lifeboats, but they were rusted in their davits.

As the *General Slocum* reached 129th Street, the flames towered ten feet above the decks. The passengers huddled at the stern of the ship in one squirming mass of humanity, and as the pressure of the crowd increased, the rails began to weaken. A few people had already climbed over the rails, ready to jump overboard if necessary.

Although Captain Van Schaick still had not sent a distress signal toward the shore, the fire was seen along each shoreline and tugs and other boats headed for the burning ship. Before long an armada of small ships was trailing the *General Slocum,* their skippers hoping to rescue some of the passengers when they jumped or fell into the river.

By 132nd Street the fire had become a huge wall of flame making its way with horrifying speed toward the people at the stern. Passengers aware, if Captain Van Schaick wasn't, that the motion of the ship was sending the flames roaring in their direction screamed for the captain to stop the ship. One man, with his dead child in his arms, made his way painfully to the bridge and burst in on Van Schaick.

"For God's sake, steer for shore!" he screamed.

Van Schaick looked at the man but did not answer.

Cursing, the man drew a pistol and fired at him, but he missed. Frustrated, he returned to the ship's stern, the burned and dead child still in his arms. Van Schaick, trembling from the experience, kept the *General Slocum* on course.

It was still a mile to North Brother Island.

The passengers at the stern faced a grim choice. They could remain on deck and hope that somehow the flames would not reach them, or they could leap overboard and trust the flotilla of ships trailing the *General Slocum* to rescue them before they drowned. Neither alternative was enticing, and most of the frightened people waited until the last minute to make up their minds.

But there were some exceptions. One woman, with clothing already ablaze, leaped overboard. Encouraged, others followed. But in this act they encountered a new danger. Many of them hit the water near the huge paddle-wheel and were sucked into the revolving blades and killed. Others found themselves whirling around helplessly in the *General Slocum*'s wake, and most of these people drowned before boats could reach them.

Near 140th Street the weakened port rail collapsed, and half the crowd still untouched by fire cascaded into the river. The rescue boats swooped down on them, pulling some from the water, but it was later estimated that they saved only one in ten.

Those left on the deck were in immediate danger of being consumed by the raging fire that had now reached midship and was rapidly approaching the last oasis on the ship. More and more of the passengers decided the river was the safest and jumped overboard. Women threw their children into the water and then leaped after them. Some jumped holding their babies in their arms. One woman dropped to the deck and gave birth prematurely to a child. With the flames approaching her, she calmly rose to her feet, picked up the newborn child, and jumped into the water. Mother and child both drowned.

As the death ship neared North Brother Island, Captain Van Schaick ordered Van Wart to steer for an area of mudflats along the shore. The ship beached with a shudder, like some stricken monster in its death throes. The trailing armada of boats quickly formed a semicircle around the stern of the fiery ship, ready to move in and rescue those who now leaped overboard. But then the final major catastrophe occurred. The entire upper deck of the paddle-wheeler collapsed, dropping its human cargo into a fiery pit below. At that moment, some four hundred people perished.

Throughout the long, agonizing death ride, as well as at its termination, individual acts of both heroism and cowardice occurred. Reverend Haas, who lost his wife and daughter in the fire, later gave this graphic view of the disaster.

> The band was playing a popular air, and the women and children were crowded around the musicians listening to the piece. The band was located aft. Then came the cry of fire and the awful scene. In three minutes—it could not have been more—all three decks were ablaze.
>
> Such scenes as followed I do not believe were ever witnessed before. Women and children were shrieking and crying in their terror of the frightful doom facing them. Some of the poor mothers had three or four little children with them, and their attempts to save the little ones were heart-rending beyond description. Everything seemed hopeless.
>
> When the fire first shot up through the hurricane deck and drove the crowd back, the panic was simply terrible. Those in the rear were swept along by the crowd in front, the women and children trying in vain to hold onto the railings and stanchions. Those who fell were crushed.
>
> My wife and daughter and myself were among those swept along in the rush. There seemed to be a general inclination to jump, and the women and children were swept over the rails like so many flies. In the terrific rush many of them were trampled on. Little children were crushed, while mothers and their babies would give wild, heart-breaking screams and then jump into the water.
>
> Soon we saw that boats from the shore were making for us, and then we had a ray of hope. It looked up to then as if no one would be saved, and with my wife and daughter I went overboard. I do not know whether we jumped or whether we were pushed over. When I rose to the top of the water I saw scores and scores of people fighting to keep afloat, and then one by one they would sink for the last time. It was awful and I was powerless to do anything. I did not see my wife or daughter.

By a great effort I managed to keep above water, but my strength was about gone when a boat picked me up.

Another minister aboard, Reverend Julius G. Schulz, said:

It is absolutely impossible to describe the horrible scene on the ship. The flames spread so rapidly that it seemed only a second before the whole craft was ablaze from end to end. Women and children jumped in the wildest manner to their deaths, while the efforts of mothers to save their little ones was the most heart-rending spectacle I have ever witnessed.

Poor Mr. Haas did his best to save his wife, but in the excitement Mrs. Haas was lost. I myself was among fifty others who were saved by a boat, the name of which I do not know.

Mary Kreuger, a single woman who suffered burns, testified:

I was on the upper deck when I was startled by a cry of fire. Then the men came along and told the women to be quiet. The advice fell on deaf ears, however, for every one became panic-stricken the minute the alarm was given.

I slid down a pole to the water and managed to get hold of a rope that was hanging alongside the ship. I had to relinquish this, however, in short order, for flames began to shoot out of the portholes right above me. Alongside of me was a little boy, and he was holding onto a life preserver. Near us was a coal barge, and a deckhand on that threw us a rope and pulled us on board the barge.

Freda Gardner, an eight-year-old child, told a story of mixed cowardice and heroism:

The first thing I knew of any trouble was when everybody started shouting and running to the back of the boat. I was knocked down but I got to my feet again. A big man stopped and put a life preserver around me. He was praying all the time and hurried to help another person. I fell again, and when I was getting up somebody—I think it was a woman—tore off my life preserver. I got to the outer rail but I was afraid to jump.

Then a man picked me up and threw me into the water. I saw him a second later swimming toward me, and it gave me courage. Then he disappeared. A plank came floating by and I grabbed it. All the time I was trying to pray. One time a man grabbed the plank and was pulling himself up on it when a woman threw her arms around his neck and the two slipped back into the water. I managed to say a prayer then and felt better. I started to pray again when a man in a rowboat reached out and pulled me in.

A man who gave his name as Miller accused the captain of gross negligence for not keeping the life preservers in a state of readiness. "I put on six of them and every one of them fell apart. I finally jumped overboard without one and swam until a tug rescued me."

Hannah Ludeman, a seventeen-year-old girl, lost her mother and two brothers in the disaster. "We were all on the upper deck at the stern of the boat," she related,

when smoke was seen and the crowd shouted that the boat was on fire. I helped mother to put on a life preserver and I got one on myself. On our part of the deck the life preservers were not burning at that time. I left mother inside a cabin and went out on the deck to see what we could do, but I had scarcely got out there when the part of the deck near where she was collapsed and she went down into the burning ship with it.

I was thrown into the water, but the life preserver held me up until I was pulled out. I don't know what happened to my brothers but I know that there was not a chance that mother could have escaped.

One sixteen-year-old girl was a heroine. She had been placed in care of a two-year-old child from another family, and she and the baby were on the hurricane deck when the fire started. Finding a life preserver in good condition under a seat, she tied it around her body and then placed the baby on her back. Then she jumped into the water and swam toward a nearby tug. She and the baby were both saved.

Minnie Weiss, eleven years old, told a heart-rending tale of tragedy.

> I started on the excursion with my mother, my brother George, and my cousin Louisa Roth. We were at the front of the boat on the second deck, except my brother who was on the upper deck, when somebody cried "Fire!" Everybody was scared and the ladies and little girls began to cry and scream and run about. My brother got to us and tried to get a life preserver for my mother and one for me, but the fire seemed to get to us right off and the life preservers were all burning. My brother finally got one that wasn't burning for me and I followed a big woman who was near me. The next thing I knew I was in the water and then this big man caught me and tied a line about me, and they pulled me on a boat. I don't know who the big woman was and I did not see my mother or brother or cousin again.

Of all the tales of tragedy that emerged from the terrible holocaust on the *General Slocum,* probably the most poignant was the strange and ironic story of six-year-old Marguerite Heims. For days no one knew whether she had burned in the fire, had drowned, or had miraculously escaped. Then one day her tiny body was found floating in the murky waters of the East River at Clinton Street within a stone's throw of the house she had lived in. Marguerite Heims had floated eight miles to come home.

The people of North Brother Island were ready when the furiously burning *General Slocum* shuddered ashore. Situated on the island was the Health Department's Isolation Hospital for patients with contagious diseases, and Dr. William Watson, head of the hospital, had all contagious disease patients locked up and guarded while he and a group of other doctors and nurses raced to the flats where the *General Slocum* was grounded. Many others had seen the fiery boat heading for the island and were prepared to try to rescue the passengers.

Tugs, rowboats, anything that would float were poised and ready. Fireboats and other fire apparatus began to play water over the ship as it beached. Across the river on the Manhattan side employees of a marble works took to boats and raced to the rescue. From all directions small boats converged on the scene.

Much of the effort was ineffectual, however. The hoses played on the ship were useless because the fire had such a head start that it was no longer possible to control it. And the heat from the conflagration was so intense that small boats had difficulty approaching the *General Slocum* close enough to make rescue attempts.

Meantime, the frantic people still aboard began to leap off the decks. Some hit the paddle wheel and were killed instantly. Some landed in water too far from the rescue boats and drowned before they could receive help. Many suffered broken necks, arms, and legs as they landed with sickening thuds on the decks of the tugs nearest the ship.

Still, there were many heroic rescues. In the course of a few minutes the tug *Wade* pulled 155 people aboard. Fireman Edward Carroll leaped from the tug and rescued eight people himself. In one case he had a sad choice to make. He grabbed three drowning children at one time and attempted to swim back to the tug with them. But he could not handle all three and had to abandon one to save the other two.

Thomas Cooney, a policeman on the island, rescued eleven people, then drowned trying to save a twelfth. Albert Rappaport, of the tug *Massasoit,* rescued five girls, two boys, and a woman. James E. Gaffney, an engineer on the island, persuaded five men to make a human chain into the water, with himself as the outermost man. Twenty people were handed back along the chain and rescued.

One tug approaching the burning ship found a pile of bodies draped in a gruesome tangle over the paddle-box. Only one person in the heap was alive—a tiny little girl named

Lizzie Krieger. She waved to the men on the tug, and they managed to pull her off the burning ship. Later she revealed that she had seen her mother burned to death. "Mother is all burned up," she cried.

The powers that some people can assume in a crisis was demonstrated by Nellie O'Donnell, a nurse from the hospital. Miss O'Donnell could not swim but became so concerned over the plight of those in the water that she waded out toward them. When the water was over her head, she started to swim. She rescued a boy and brought him ashore. Dripping wet, she stood for a moment on the shore, amazed at her own accomplishment. Then she went back in and rescued nine more people before she became exhausted and sank wearily on the shore.

A Dr. McLaughlin, who was in charge of the tuberculosis patients at the hospital, commandeered a small rowboat and set out for the burning excursion boat. He quickly rescued six people and brought them ashore, then looked after a young boy lying on the mudflats who needed attention.

George W. Johnston, who had come to North Brother Island that day to spend time with his friend, James Owen, found himself close to the area where the *General Slocum* beached. He and Owen set out for the ship in a large rowboat and came in close to her stern. They shouted for the people to jump, and when they hesitated, Johnston, a strong swimmer, went into the water to encourage them. Some of the passengers then jumped, and within a few minutes Johnston had grabbed them and pulled them to the boat, where Owen hauled them aboard.

Owen rowed ashore with eighteen people piled like cordwood in the boat. Meantime, Johnston was in the water with a woman clinging to his back and a small boy hanging onto his left arm. He managed to keep both of his charges above water until Owen came back with the boat. Then he turned and picked up two more children. Finally exhausted,

Johnston climbed into the boat and went ashore with Owen and their four charges. By the time they landed, Johnston was ready to row out again, but this time could find no one alive. Johnston described the scene of horror thus:

> It was simply awful. Our boat was drifting and we were impeded by the dead, our oars striking them everywhere. Many times we would run into the floating bodies. The worst sight was the stern of the boat where there was a small boy burning and in some way caught so that he could not free himself. He was quite dead, but close to him was a live fellow with long golden hair. His hair was on fire and he was doing his best to beat it out. The heat was blistering but we tried to get to him.
>
> We were still some distance off when he fell back into the flames. I was almost finished myself with what I had seen and the struggle in the water. Poor Owen was crying like a baby and the two children in the bottom of the boat [Owen and Johnston had decided not to leave them on shore alone] were shivering and pleading for their mothers. We couldn't be of any more aid, and it did not seem worthwhile to try to pick up the dead, so we rowed back through them to the shore.

Within ten minutes of the time the *General Slocum* was beached on North Brother Island, no one aboard the ship was alive. The ship was a mass of fire from prow to stern, with flames shooting into the air as high as sixty feet. Finally, as if wanting to get away from the scene of its own destruction, the burned out *General Slocum* freed itself from the mudflats and began drifting north. Near Hunts Point she sank, taking with her hundreds of burned and charred corpses.

And what of Captain Van Schaick? When the excursion boat beached, he ran to the stern and leaped into the water. His face had been burned and one eye damaged, and when he leaped into the water, he split his heel on a submerged rock.

At the hospital, he told three different stories to a reporter for the *New York Times*. Asked why he had insisted on continuing to North Brother Island instead of putting into shore sooner, he said, "I tried to put into the New York shore but was warned off." This assertion was denied by scores of people who had watched the headlong progress of the boat up the East River.

The captain also stated that it was not over four minutes from the time he first heard of the fire aboard until he beached the vessel. When it was pointed out that this was an impossibility, the captain said, "Well, it might have been longer, but it did not seem longer to me."

Van Schaick next claimed that the fire hose aboard the ship worked, later contradicting himself by saying the hose had been torn from the hands of his men while they were trying to bring the water into play. Actually, the hose, which had been cheaply made of linen, had split down its length.

Finally, shaken by the disaster and the interrogation, he refused to answer any more questions.

By midnight six hundred bodies were stretched in long rows on the beach of North Brother Island. Some four hundred more were estimated to be dead on the ship itself or in the waters. The death count never was determined exactly, since many bodies were carried away by the river and others were given up by the river long after the tragedy occurred. The U.S. Marine Inspection Department finally set the dead at 938; the New York Police Department count was 1,031. The *World Almanac* sets the number killed at 1,030.

In any case, almost all of the casualties were members of St. Mark's parish. In one apartment house where fifteen families had lived, there remained fifteen wifeless and childrenless fathers. A total of one hundred twenty men lost their entire families.

Local and federal investigations revealed neglect and ineptness by the ship's owners, the captain, and the crew. It was found that the life preservers on the *General Slocum* had been wired in place when the vessel was built thirteen years before and had never been inspected since. In fact, it was discovered that the manufacturer of the life preservers had actually inserted seven-inch cast iron bars in each cork block to bring the preservers up to the required weight!

Henry Lundberg, an inspector for the marine inspection service, had "gone over" the ship the day before the tragedy and had approved it, but under questioning he admitted he had not checked the life preservers, lifeboats, or fire hoses.

When the fire started, the crew had shown complete inability to cope with it, since the ship's personnel included men from various walks of life who were unfamiliar with marine practices. None had ever participated in a fire drill or been told what to do in such an emergency. It was also discovered that First Mate Ed Flanagan, although he had been on the *General Slocum* for two years, had never even been licensed.

The precise cause of the fire was never determined. One explanation was that a porter, filling and lighting lamps in the storage room, had dropped a match in the hay scattered about. Another theory was that a workman with a lighted torch passing through the storage room had inadvertently started the fire. Still another story located the origin of the fire in the galley near the storage room where chowder was being cooked.

In any event, a coroner's jury investigating the *General Slocum* disaster promptly decreed that eleven men be held under bond on manslaughter charges. They included the president, secretary, and five shareholders of the Knickerbocker Steamboat Company, owner of the *General Slocum;* steamboat inspector Henry Lundberg, who had made such a cursory inspection of the vessel; the commodore of the Knickerbocker Fleet; First Mate Flanagan; and Captain William Van Schaick.

Almost at once President Theodore Roosevelt demanded a complete inspection of all excursion boats and a thorough investigation of the steamboat inspection service, after which regulations were tightened. But one of the most amazing miscarriages of justice in marine history occurred when charges were dropped against all of those held—with the exception of Captain Van Schaick.

The captain did not go on trial until January, 1906. In his defense he cited his forty years as a safe navigator and pointed out that he had not left his ship until it was grounded and he could be of no more help. He had not beached the *General Slocum* sooner because of shallow water in some places and gas storage tanks along the shores, he said. The condition of the fire hoses, lifeboats, and life preservers was impossible for him to explain.

Captain Van Schaick was sentenced to ten years in prison for manslaughter. On two occasions his wife petitioned President Roosevelt for his pardon, but the President denied the request. In 1912, after spending six years in prison, Van Schaick was pardoned by President William Howard Taft. He died in 1927 at the age of ninety.

Miraculously, the *General Slocum* herself was one of the survivors, although her new life was considerably reduced in stature. Her blackened hulk was raised and became the barge *Maryland*. Eventually, though, she came to a violent end. In 1912, on a trip from Camden to Newark, she foundered in a gale off the New Jersey coast. Federal engineers decided she was a hazard to shipping and destroyed her with dynamite.

5

Double Trouble in San Francisco (1906)

THE year: 1906. The place: San Francisco. A flourishing and prosperous city, created not in the hasty manner of a boom town but by the kind of steady progress that brings stability and character to a growing metropolis. One of the world's young cities—as compared to New York, London or Paris—feeling its oats a little, knowing it is going places, sure of its future. A thriving town of four hundred thousand people, with differing cultures, dissimilar ways. A city divided, like Gaul, into three parts: the elegant mansions of the wealthy on Nob Hill; the tenements of the poor, south of Market Street; and the vile dens of the Barbary Coast, which had already earned the town a scarlet reputation for its sin, its vice, its wickedness. In short, a city of great promise.

This was San Francisco, a throbbing, energetic metropolis whose vigor and ability to rise from disaster would be tested, in 1906, in a crucible of earthquake and fire that would send it reeling to its lowest ebb.

On the evening of April 17, 1906, Enrico Caruso, the

famed Italian operatic tenor, returned to his ornate suite in San Francisco's Palace Hotel a pleased and proud man. His performance as Don José in Georges Bizet's *Carmen* had won him a standing ovation from some three thousand tastefully dressed people at San Francisco's Grand Opera House. Afterward, he had dined sumptuously at a famous restaurant, the final touch to a magnificent and thoroughly satisfying evening.

Caruso was certain that a Supreme Being somewhere was watching over him with extraordinary care. Ten days ago Mount Vesuvius had erupted, threatening several nearby towns including Naples; Caruso, who was genuinely frightened by such capers of nature, had just left Naples on a journey to the United States, where he was scheduled to tour with the New York Metropolitan Opera Company. In Caruso's mind, he had been saved from terrible destruction in Naples and transplanted to much-safer San Francisco so that he could add glory to his already illustrious name—and all this was assuredly the work of Someone on high.

There was another performer in San Francisco that night, not so prominent as Caruso but an actor who was nonetheless on his way up the theatrical ladder. John Barrymore, then twenty-four years old, was soon to board a ship for Australia, where he was slated to appear in several plays. Barrymore had attended the performance of *Carmen,* and afterward, one story claims, he took the winsome young lady who had sat next to him at the opera to dinner, spending the entire night with her. He is reported to have appeared on the streets the following morning still dressed in his top hat and opera cape. Another story has it that after the opera Barrymore met a young man who boasted about his collection of fine Chinese porcelain, and that Barrymore was interested enough to go to his apartment to see it. There he spent the night, sleeping on a divan.

There were other famous people in town that night: Olive

Fremstad, who played the title role opposite Caruso in *Carmen;* Jack London, the novelist; author Mary Austin; and, just north of the city in Sebastopol, Luther Burbank, the horticulturist—all fated to live through a catastrophe that no one anticipated on the warm, fogless, star-studded evening of April 17, 1906.

One man of lesser renown, though his name was of some importance to the City of San Francisco, was Dennis Sullivan, the town's fire chief—the only man in the city who had formulated detailed plans to save San Francisco in the event of a great conflagration. He had just returned from an exhausting all-night ordeal fighting a fire at the California Cannery Company warehouse. Unlike those who had spent the night in revelry, Sullivan went to bed immediately in his room on the third floor of the Bush Street firehouse. Dawn was only a couple of hours away when he retired.

At 5:12 on Wednesday morning, April 18, the earthquake hit. It shook the city savagely, toppling houses, collapsing chimneys, reducing huge buildings to rubble, opening fissures in the streets—all in a matter of minutes.

The northern half of the San Andreas Fault, a geologically ancient fracture line in the earth's crust, had slipped its moorings. The result was one of the most destructive earthquakes in American history. It ravished not only San Francisco but an area 300 miles long up and down the coast and for as much as 40 miles inland. Its awesome power split giant redwood trees down the middle, reducing them to kindling. Highways were torn up, houses and barns demolished, the ground itself churned and uprooted.

The quake awakened sleepers from Ukiah in the north to San Luis Obispo in the south. It wreaked destruction on large towns and small. Every brick building in Santa Rosa collapsed. The business section of San Jose became instant rubble. The insane asylum at Agnews crashed down on its patients. More than one hundred were killed; others escaped to roam the countryside. In Solano County, northeast of San Francisco, a mile of railroad track sank from three to six feet. Stanford University, founded only fifteen years before, suffered severe damage to its campus.

Perhaps the coolest man to be jarred out of his bed by the early-morning quake was Professor George Davidson, of the University of California in Oakland, who had formerly been connected with the United States Geodetic Society. Even as the shock waves rolled beneath him, he was able to record for posterity the sequence of the earth's giant shudder. Later he explained it this way:

> The earthquake came from north to south and the only description I am able to give of its effect is that it seemed like a terrier shaking a rat. I was in bed, but was awakened at the first shock. I began to count the seconds as I went toward the table where my watch was, being able through my practice closely to approximate the time in that manner. The shock came at 5:12 o'clock. The first sixty seconds was the most severe. From that time on it decreased gradually for about thirty seconds.
>
> There was then the slightest perceptible lull. Then the shock continued for sixty seconds longer, being slighter in degree this time. There were two more slight shocks afterward, which I did not time. Then at 5:14 I recorded a shock of five seconds duration and one at 5:15 of two seconds. There were slight shocks at 5:17 and 5:27. At 6:50 there was a sharp shock of several seconds.

Enrico Caruso, exhibiting the high-strung emotions common to many artists, was not nearly so calm as Professor Davidson. The singer was jolted out of a sound sleep to find his palatial suite dancing crazily about him. Plaster was raining from the ceiling, powdering the room with a white dust. Bureau drawers were scattered in haphazard fashion. Pictures had tumbled from the walls. Everything was in disarray.

The trembling Caruso burst into indignant tears at the sight. Whereas the previous night he had believed that a

Supreme Being watched over him, he was now equally convinced that the Great One had deserted him. He was still in his bed, glossy tears streaming down his face, when Alfred Hertz, conductor of the Grand Opera House orchestra, rushed in.

The frantic Caruso cried out that he was doomed and that the shock had ruined his vocal cords.

Hertz urged Caruso to look out of the window. He gazed down on a scene of great confusion. Nearby buildings had collapsed. Fronts were ripped off apartments. People, some only half-dressed, ran about wildly. Life was clearly not over for everyone, and especially not for Caruso.

And when Hertz demanded that he sing, to test his voice, and he was able to do so, Caruso discovered also that his powerful tones somehow seemed to soothe the people in the street.

Later Caruso described his experience thus:

> I waked up feeling my bed rocking as though I am on a ship. From the window I see buildings shaking, big pieces of masonry falling. I hear the screams of men and women and children. The ceiling plaster fell in a great shower. I run into the street. All day I wander about. I try to get away, but soldiers will not let me pass. That night I sleep on the hard ground. My legs ache yet from so rough a bed.

In fact, the day after the quake the Metropolitan Opera troupe managed, for $300, to hire a wagon and driver to take them and their luggage to the ferry to Oakland. It was said that Caruso, in a defiant gesture as the ferry departed, shook his fist at the city of San Francisco and vowed never to return.

John Barrymore was less ruffled by the earthquake. Wherever and however he spent the night, he nevertheless appeared on the rubble-strewn streets at ten o'clock—five hours after the quake hit—still dressed for the opera. Later that day, learning that he had been reported missing, he decided to send a message denying the report to his sister,

Ethel, in New York. He persuaded a newspaperman sending a dispatch about the earthquake to New York to include a message to his sister. In essence, the message conveyed the information that he had been awakened by the quake, had wandered around the street for hours in a semi-daze, and finally had been conscripted by a brigade of soldiers who had put a shovel in his unaccustomed hands and forced him to work clearing rubble from the streets.

Whether this actually happened or not is open to question. Barrymore, like most actors, was not above stretching the truth for dramatic purposes. It is true, however, that able-bodied men were compelled to do such emergency work, and the youthful Barrymore might well have looked capable of wielding a shovel.

When Ethel Barrymore received the message, she showed it to her uncle, John Drew, and asked him if he believed it. Drew, who was not enchanted by young Barrymore's efforts to become an actor, then made the remark that became almost as memorable as the earthquake itself.

"I certainly do believe it," he said. "It *would* take an act of God to get John out of bed and the U.S. Army to put him to work!"

Luther Burbank, in Sebastopol, was awakened when his bed began to rock like a cradle. However, he remained calm, later explaining that he had been through more than a hundred earthquakes in his time. He rose from his bed, staggered over a moving floor, and went into the yard to inspect his beloved plants—his first concern when the earth began to roll. He found that though they had been moved out of position by the quake, they had not been seriously damaged. He breathed a sigh of relief and went back into the house.

On the third floor of the Bush Street firehouse, Fire Chief Dennis Sullivan leaped from his bed when the quake struck. His greatest concern was his wife, who was in a bedroom down the hall. He ran down the corridor to rescue her, but he never reached his destination. Next door to the firehouse the California Hotel swayed like a reed in the wind and suddenly collapsed with a roar. A heavy slab of cornice fell through the roof of the firehouse and pinned Sullivan in the wreckage. His wife was unhurt, but a few days later Sullivan died of his injuries.

San Francisco, perilously situated on a peninsula with the Pacific Ocean on the west and San Francisco Bay on the east, was shaken four ways by the monster quake. The first sign of trouble was an ominous roar. Next the city was shaken by a horizontal shifting in the San Andreas Fault, a dislocation that caused the earth under the city to roll like waves in the sea. Almost at the same time there was a vertical movement, causing the needles of seismographs to rotate in circles and the city to reel under the violence of the four-way attack.

Tall, steel-framed buildings rocked back and forth, but most of these stood. Other buildings, less well constructed, were reduced to shambles. The City Hall, a $7,000,000 edifice, collapsed. The Hall of Justice and the jail were both badly damaged. A printing plant collapsed, dropping the heavy presses through the weakened floors. People were killed by falling bricks and cornices as they staggered down the heaving streets.

On Market Street, a major avenue, the streetcar tracks were uprooted and twisted into surrealistic shapes. Chimneys all over the city toppled or were thrown from roofs. Underground water pipes were crushed, and water gurgled to the surface and ran down the streets. On the bay, Long Wharf disintegrated, dumping thousands of tons of coal into the waters. Chinatown was shattered. In the slums along Howard Street, flophouses and decrepit hotels were leveled, killing hundreds of indigents. In the wholesale district men, horses, wagons, and bricks from crumbling walls were mixed in an indescribable heap on the streets. Only the fact that the quake struck at 5:12 in the morning—before factory and dock workers had arrived on their jobs and children at their schools—kept the death toll from mounting astronomically.

But this epic seismic disturbance was only the beginning of an even greater tragedy. The earthquake was the lighted match that ignited the greatest fire in the history of this country—a fire that demolished the greater part of San Francisco. Before it was over, a huge circle twenty-five miles in circumference—encompassing the business district, industrial area and residential sections—was burned to a crisp.

Most San Franciscans at the time insisted that the earthquake was "only a little worse" than others experienced, in 1864, 1898, and 1900. But the fire that followed, in their opinion, was a living hell. The fire did not begin in one area and sweep through the city, thus giving firefighters a chance to stop it somewhere along the way. Instead, it sprang up in a thousand different places at one time, making control almost impossible.

The scattered fires were ignited by the general disruption caused by the quake. Lamps and gasoline stoves were upset. House chimneys were ruptured. Gas mains were broken. Trolley wires lay in the streets, spitting sparks. Furnaces were upset, and industrial chemicals were scattered about. Boiler rooms in factories were wrecked, causing the boilers to burst. And when the gas works near Market Street blew up, the stage was set for total disaster.

Minutes after the quake struck, the San Francisco Fire Department received sixteen calls. Equipment was rushed from one emergency to another, but there was little the firemen could do. Many private homes were ablaze, and some of these fires were extinguished, but the fire roared out of control in the factory and wholesale centers. Even under the best of conditions the firefighters would have been hard pressed to conquer the flames. But in this case it was hopeless. The loss of Chief Sullivan really had no effect. The quake had broken open the water mains, and there was no water with which to fight the conflagration.

Within three hours the many small fires had merged into bigger ones, and nine distinct major blazes were raging in the city. There was only a slight wind, but it was enough to spread the holocaust. The intense heat from burning buildings ignited those across the street by spontaneous combustion. Oil tanks, freight sheds, warehouses, factories, lumber yards, retail stores, and jerry-built homes fed the fire.

In the burning streets people fled in mass panic. Many tried to save their most precious belongings, carrying them in wheelbarrows, baby carriages, and wagons. One woman fled with a canary in a cage. Another had a clock under her arm. Some carried beloved pictures or treasured vases. A couple pushed a sewing machine down the street. All were in haste, fleeing the fiery destruction that was closing in on them.

One man was cool enough to take his time. With his neighbors' houses burning around him, he stood before a mirror shaving himself and combing his hair. Then he dressed, packed a bag, and strolled from his home. It was not until he was in the street that he noticed he had neglected to put on his pants. Whether or not he returned to his house to get them is not of record.

The crowded district of the poor, south of Market Street, was destroyed within hours. The Palace Hotel, on Market, where Enrico Caruso had stayed, was gutted. A half-block away the Call Newspaper Building became a giant torch.

All through that first day the exodus from the city continued. Some people were satisfied merely to reach open ground on the city's edges. Others headed for the waterfront, hoping to board ferries that would take them across the bay. These latter people were fortunate, since it was in the bay area that the firefighters enjoyed a small victory. Using fireboats to spray the shoreline with water from the bay, the firemen were able to keep most of the wharves and docks from burning.

Everything that floated—ferries, barges, tugboats—was used to take people out of the burning city. Many families who fled toward open country slept the first night in Golden Gate Park.

Even while San Francisco burned, attempts to bring the city to order were being made. Mayor Eugene Schmitz and General Frederick A. Funston had joined hands to save what they could of the fire-swept city. Both were rugged men. Schmitz, of German-Irish extraction, was a veteran of the Yukon Gold Rush as well as a talented violinist who had risen from the Musicians Union to the mayor's chair. Funston, formerly a Kansas farmer, was an army man in command of what was then called the California Department of the Army. These two men early in the disaster formed a partnership of action.

The men under Funston's command were sent into the stricken area to fight fires, organize hospital and first-aid units in the rubble of the streets, form rescue parties, and guard against looters. At the same time the mayor wired other communities for help, and within hours bedding, tents, and food were beginning to reach the city. Schmitz also conscripted every able-bodied man for firefighting and rescue work and placed them under the command of Funston. As the fire grew in intensity, Schmitz sent horses and wagons to powder houses in outlying areas with orders to bring all the dynamite available in the event blowing up buildings was necessary to halt the fire.

But the great fire raged all day long and through the night. On Thursday, April 19, it leaped the barrier of Market Street to seek additional victims in the business, hotel, and amusement areas. What the earthquake left of Chinatown was quickly burned out; its dope dens were destroyed as addicts lay half-conscious in their cribs. The notorious Barbary Coast went up in lurid flames. A section of Italians, on Telegraph Hill, used casks of precious wine to soak rags and put out embers that fell on their roofs. On Nob Hill the wealthy set prepared to abandon their luxurious mansions to the approaching fire.

One major problem facing the army and the men conscripted to help were victims pinned under debris by the earthquake. Men worked frantically to free them before the purgatory of fire reached them. Some were rescued in time; others perished as the flames swept over them. In one case, a man pinned down by wreckage as a dozen men worked feverishly to free him sensed that the task was impossible and begged the soldiers to shoot him. The men worked until they, themselves, were burned about their hands and faces. Then, realizing the hopelessness of the situation, one of the workers took a gun from his belt and shot the man.

In another area five men were dragged from the ruins of a collapsed building and laid in the street. Three priests from St. Patrick's Church granted them the last rites as a large piece of coping overhead threatened to fall and crush the priests to death. When the ceremony was completed, all eight men were moved to a safer place. Three of the victims later died.

Children suffered as well. Mrs. Henry Huskey, a San Francisco resident, described the plight of the children thus:

> Under our own observation was the case of a child ill of diphtheria who was carried into the streets Wednesday night by her parents and died in agony on a lawn the next morning. Utter lack of water in some districts got the children moaning and pleading for a drink. Men of ruined families made every human effort to satisfy the thirst of their little ones. At last, in desperation, they invaded the neighborhood saloons and brought whiskey to their women.
>
> Unable longer to withstand the pleading of their children, mothers poured small quantities of the fiery liquor into tin cans and other receptacles and gave it to the tots to drink. The natural result was to increase the pangs of thirst twentyfold, and the sight of the woebegone, staggering children was then witnessed.

O. M. Nichols, a New Yorker visiting San Francisco, told what it cost him to escape the city:

> Reaching the bay, we found there was no ferry. An old fellow had a tugboat tied up. There were several of us wanting to go to Oakland. We asked the boatman what he would take to land us there. He said $250. We chipped in $50 apiece and took the tug.

George Musgrove, a theatrical manager from Australia, described graphically the panic at dockside:

> I shall never forget the scene at the ferry house. It was bedlam, pandemonium and hell rolled into one big pile. There must have been 10,000 persons trying to get on a boat. Men and women fought like wildcats to push their way aboard. Clothes were torn from the backs of both men and women indiscriminately. Women fainted and there was no water at hand with which to revive them.
>
> Men lost their reason at those awful moments. One big, strong man beat his head against one of the iron pillars on the dock and cried out in a loud voice, "This fire must be put out! The city must be saved."
>
> When the gates were opened to the boat the mad rush began. All were swept aboard in an irresistible tide. We were jammed on the deck like sardines in a box. No one cared. We were out of the smoke-filled atmosphere and were on our way to a place of comparative safety.
>
> Almost everyone thought that the end of the world was at hand. Native sons of San Francisco had experienced earthquakes before. But this was something worse. They had no communication with the outside world. Naturally they imagined that the disturbances were being repeated all over the country.

Throughout the burning city wild rumors were spreading: it was a worldwide disaster; New York City had been engulfed by a great tidal wave; Chicago had slipped into Lake Michigan; London and Paris had been leveled by fire. Gruesome stories emerged: the quake had freed wild animals at the zoo, and they were now feeding on people who had fled to Golden Gate Park; the quake had churned up buried bodies, scattering putrid corpses among those seeking safety from the fire in the cemeteries. None of these rumors was true, but they added a macabre horror to the general scene.

Miss Martha Sibbals, who escaped from the tottering Randolph Hotel during the quake, told another harrowing tale:

> Someone passing us in the street advised us to get to as high ground as possible. We started walking as fast as possible to the high parks back of the city. Fire was starting in hundreds of places over the city and the streets were crowded with hurrying refugees.
>
> Where people were unable to procure horses, men and women had harnessed themselves to carriages and were drawing their belongings over the streets. In the residential districts where wealthy people lived we saw automobiles drawn up and loaded down before houses. The owners remained until the flames came too near, and then they got into their machines and made for the hills. We saw one man pay $2,000 for an automobile in which to take his family to a place of safety.
>
> Before night we reached high ground away from the flames. People half-clothed, unfed, hysterical, searching for loved ones, crowded the ground. We passed the night sleepless with a panic-stricken multitude.
>
> In the morning we started toward the harbor with the assistance of soldiers from the Presidio, who had already been on duty twenty-four hours. We got to the wharf and sought to get a launch to Oakland. We were unable to do so, but we were kindly treated by an old skipper who was himself in deep grief because his brother had been crushed to death in their little house. He gave us coffee, the only nourishment we had had except for a few crackers in twenty-four hours.
>
> Then the skipper saw the Government boat cruising in the bay, and said if we could reach the Presidio wharf we could escape on the Government boat. We therefore hurried toward the Presidio, greatly impeded by fissures which stretched long distances and around which we had to make our way. At the Presidio we were taken aboard with other refugees, and a short time later we were safe in Oakland.

Mr. C. C. Kendall, of Omaha, described his attempt to escape the burning city:

> I climbed over dead bodies, picked my way around flaming debris, and went over almost insurmountable obstacles to get out of San Francisco. The debris was piled up along Market Street. Fires were raging in every direction. Market Street in my location had sunk at least four feet. It is only a few blocks from the Palace to the ferry, but it took me from 6:00 A.M. to 10:15 A.M. to cover the distance.
>
> Men and women fought each other at the entrance to the ferry like infuriated animals. As the boat pulled out over the bay, the smoke and flame in the city rose sky-high, and the roar of falling buildings and the cries of the people filled the air.

One man experienced in both earthquakes and fires, Dr. Frank A. Brewster, visiting from New York, described his experience this way:

> I witnessed the burning of Chicago in 1871 and was in Charleston when an earthquake wreaked great destruction there, but the San Francisco horror far outshadowed those calamities. I cannot believe the loss of life in San Francisco was confined to mere hundreds. I would be the last to stretch the facts, but I am convinced several thousand persons suffered death.
>
> Thieves and ghouls were dealt with summarily. I witnessed the demise of several ghoulish men. I saw a fellow cutting rings from the hands of a dead woman. There was a rush to the scene of several men and within a few minutes the robber was dangling from a pole.
>
> One innocent man met his death at the hands of the military. He was a cashier of a bank and refused to obey a command to halt, but continued to run into the bank building. He was shot.

Miss Artie Hall, a member of a family of jugglers appearing at the San Francisco Orpheum, attempted to retrieve some equipment in her dressing room on the afternoon of the earthquake. The building was already on fire, and as she entered the building, it collapsed and crushed her to death.

As always occurs, some people made money out of the tragedy. Arthur Woodson, a Chicagoan, was staying at the Pacific Hotel at the time the earthquake struck. Later he said:

> After breakfast I hustled around to get over the bay to Oakland. A few hacks were in commission and a regular auction was held over the seats sold to each customer. The prices ranged from $20 to $100. I got a seat in one of the hacks for $35. There were three other men in it who paid $50 for seats.
>
> We had to go to the ferry in a roundabout way, and when we had covered several blocks two men halted the driver. They offered him $300 apiece for the privilege of riding to the ferry. The driver took one up on the box with him, handed me my $35, and made me get out so that the other $300 passenger could get into the carriage.
>
> I hailed the next hack that came along and got up on the box with the driver. Another fellow from the sidewalk called out that he would give $75 to be taken to the ferry, but I told the hack driver that I would throw him off the box if he stopped as he had no room, inside or out, for another passsenger.

As thousands of weary refugees spent the second night in Golden Gate Park, watching the inferno devour their city, the fire swept toward the western section of town. The relentless flames played no favorites. They destroyed the wooden residences on the western slopes of the town and also burned to cinders the wealthy mansions on Nob Hill.

But one large residential section in the western section of town was still untouched, and firefighters saw a chance to save it. At Van Ness Avenue, one of the broadest thoroughfares in the city, a last desperate stand was made. Using dynamite brought in from outlying areas, a mile-long row of magnificent houses was ruthlessly blown up. With the merciful aid of a slight wind out of the west, the fire was halted at that point, having burned for three days and nights. It had already been stopped on the east by the bay and on the south by the Southern Pacific railroad yards.

When it was finally over, the destruction was assessed. It was decided that the earthquake caused only 15 percent of the damage; the fire ravaged the rest of the town. It went down in history as the world's worst conflagration; the fire and the earthquake together were considered the most costly double disaster ever visited on a city.

The fire had completely destroyed four hundred ninety city blocks and had partially ruined thirty-two others. More than twenty-eight thousand buildings were burned out. Property loss was estimated at $500,000,000. The death toll was approximately four hundred fifty with many others injured.

San Francisco's recovery was remarkably fast. Civic committees worked on such problems as relief, finance, reconstruction, and transportation. Cities all across the country sent aid, as did the U. S. Congress and wealthy persons with sympathy for the stricken city.

Rebuilding started immediately. A new and more effective water system was devised, with separate systems for public use and for fighting future fires. Around the city reservoirs were built, and more effective systems for pumping salt water from the bay into the city were designed.

Only six days after the quake and fire the first contract for a new building was signed. Lumber ships steamed into the harbor, and rebuilding of houses commenced.

Within three years the city was completely restored, a city to which the earthquake and fire had only been a temporary step backward. The City on the Bay had lost a battle but had won a war. It was still sure of its future.

6

New York's Triangle Tragedy (1911)

FRANTIC and terrified, the women began to jump from the burning building on the northwest corner of Greene Street and Washington Place even before the New York Fire Department arrived on the scene. Screaming as they fell, they struck the pavement and died instantly. Firemen found it almost impossible to use their hoses, ladders and other equipment because of the twisted pile of bodies on the sidewalk. They had arrived within minutes of the first alarm, but it was already too late.

Max Blanck, a short, stocky man with a keen business mind, sat in his office on the top floor of the ten-story Asch Building and contemplated the success of the business he and his partner, Isaac Harris, had fashioned for themselves. These two men ran the Triangle Shirtwaist Company, which occupied the top three floors. As its name implied, the company made tailored blouses for women, and it was one of the largest and most successful shops in the city. Its one thousand workers, mostly women, worked mainly on the eighth and ninth floors. On the tenth floor were the administrative offices.

The year was 1911. The Triangle Shirtwaist Company was a typical "loft factory" of the time. Because of a shortage of appropriate factory space in New York City during the early part of the twentieth century, the top floors of existing office buildings were often utilized for small manufacturing operations. New York had almost 800 such factories.

There was good reason for situating these factories on the top floors, if in the quest for dollars one disregarded safety; New York factory regulations called for two hundred fifty cubic feet of air for each worker. Loft buildings usually had eleven-foot ceilings, which met the standards even if you packed the rooms with workers. Besides, electricity bills could be cut because there was always plenty of natural light on the top floors.

Working conditions in the lofts were generally poor, and this was true of the Triangle Shirtwaist Company. On the eighth and ninth floors, young women at sewing machines worked in long lines, elbow to elbow. The backs of the chairs on one line touched the backs of those on the next line, making it difficult to move about. A few men, called cutters, worked at long tables nearby. The overcrowded, sweatshop conditions represented a direct invitation to disaster.

If Max Blanck was aware of the danger, he chose to ignore it. On this particular day, in fact, he was a very contented man. He had just weathered a strike of the Waistmakers Union by the simple expedient of hiring the non-union employees who now were keeping the place humming. Most of the workers were young—between fifteen and twenty-five years old—and almost all of them were immigrants who could speak little English. He paid them from five to seven dollars a week, depending on how hard and long they were willing to work.

On the fateful Saturday that fire struck, only slightly

more than half his employees were there—about six hundred in all. They were working overtime to catch up on back orders that had accumulated during the strike. The progress report from Blanck's factory manager, Max Bernstein, was good. The backlog was being rapidly filled.

The afternoon was waning—it was now 4:30 P.M.—and two of Blanck's six children were in his office with their French governess, Mlle. Ehresmann. His wife and four other children were vacationing in Florida. He planned to leave the building in a few minutes and perhaps buy the two children—Henrietta, thirteen, and Mildred, five—a special treat on the way home.

On the eighth and ninth floors the long work day also was drawing to an end, and the women chattered in various tongues as they prepared to go home. Each had received an envelope containing a week's pay. It was not much, but in many cases she was the sole support of her family, and it had to do.

Factory manager Bernstein and Max Rother, a tailor who also served as a foreman, were on the eighth floor when the fire broke out. As usual, it was their duty not only to see that the women left the building in an orderly manner, but also that the factory was in readiness for the next day. They had already stocked it with new material for a quick resumption of overtime work on Sunday, but the debris left from this day's labor had not yet been cleaned up.

In the narrow aisles were baskets stacked with cut goods of lace and silk. On the cutting tables layers of linen and cotton fabric were piled high, ready for instant use. Huge bins were filled with scrap and waste material, and the floor was littered with remnants of cloth. On overhead lines, finished shirtwaists were hung as if on display. Although the Asch Building was classified as fireproof, the material that cluttered the Triangle Shirtwaist Company was highly combustible.

Of course, it was no secret to the management that the

materials represented a fire hazard. In fact, the company had taken a few steps in the direction of fire protection. Along the walls were buckets of water that had been used on several occasions to put out small fires. There was also a fire hose and a valve wheel, unused for a long time. The "No Smoking" rule was generally ignored by the men who worked as cutters at the tables. They had become ingeniously talented at cupping a cigarette in the palm of one hand and taking surreptitious puffs, and although the management was aware of this devious flouting of the rules there was a reluctance to enforce the order. Management took the position that if you hoped to get any work done at all, you had to allow a man a modicum of pleasure.

Even more dangerous than the cluttered working area and the clandestine smoking was the fact that exits from the building were criminally inadequate. There were two elevators—one of which was usually inoperative—and access to them was down a hallway so narrow that the workers leaving the floor had to pass through in single file. This was ideal from the company's point of view, because it was easy for inspectors to examine the women's handbags as they left. The Triangle Shirtwaist Company made certain that its employees did not steal any of the attractively finished shirtwaists.

For those hardy enough to want to walk down eight flights of steps, there were two stairways leading to ground level. One emerged on Greene Street, the other on Washington Place. Actually, only one of these could be used; the Washington Place doors on the eighth and ninth floors were bolted shut and could not be opened from the inside. This was to keep workers from wasting company time by slipping outside for a few moments of rest. The Greene Street door on the eighth floor could be used at quitting time, but the frightening flaw in this exit was that the door opened inwards, thus making it difficult to use in case of a solid rush of people for the doorway.

The only other exit was a decrepit iron fire escape leading

to an enclosed courtyard. It was so rickety that it seemed ready to collapse under the weight of only a few people. Just a foot and a half wide, it could accommodate only persons walking in single file, and it was later estimated that it would have taken three hours for those working on the top three floors to descend by this route.

New York Fire Chief Edward Croker was well aware of the fire hazards that existed in all loft factories, and he had urged city officials to take action. Frustrated in his attempts to awaken them from their apathy, Croker warned the owners of the loft factories—including Max Blanck—to hold periodic fire drills. But in Blanck's case, at least, the warning was ignored.

Thus the Triangle Shirtwaist Company was ripe for disaster. Had some firebug deliberately prepared the company for a fire, he could not have done a better job. Highly ignitable material strewn around haphazardly, a meager water supply, clandestine smokers, and inadequate exits combined to make tragedy a certainty. The only question was when.

The inevitable moment arrived at 4:40 P.M. on March 25, 1911. The exact cause of the fire is unknown. Various possibilities have surfaced. A cutter may have carelessly tossed away a match and set some waste afire. A smoker may have dropped a cigarette. A cigarette, a match, or a spark from an electric motor could have ignited gasoline used to heat pressing irons. Whatever the cause, the fire started in a rag bin on the Greene Street side of the eighth floor. It was a tiny flame, easily extinguishable had it been caught in time. But no one noticed it at first. Then one of the women screamed that frightening word that almost always results in panic:

"Fire!"

Manager Bernstein and Rother were on the Washington Place side of the loft when they heard the warning cry. Both grabbed buckets of water and raced toward the fire. But by the time they reached the scene, the fire had gained so much momentum that the two buckets of water failed to extinguish it. A couple of men then raced to the fire hose. While one man rolled out the hose, the other struggled to turn the valve. But the valve wheel was so rusty that it would not budge, and the hose itself had practically rotted away.

Meanwhile, the fire blazed up and spread to a string of shirtwaists hanging on an overhead line. The tragic drama had begun.

By this time the women, responding to the outcry, had risen to their feet in panic. Because the rows of chairs were so close together, there was great confusion as the workers attempted to get out of the narrow aisleways. The fire leaped to a cutting table and ignited rolls of flimsy material there. Thick smoke began to permeate the room. Men began to shout instructions, trying to calm the women, but they were already screaming in terror and stampeding for the exits.

In desperation the workers fled in all directions, each trying to reach one of the four possible escape routes—the staircases descending to Greene Street and Washington Place, the almost unusable fire escape, and—on this particular day—the one elevator that was running. Most of the women headed for the elevator, not realizing that it could hold only twelve at one time and that the fire was spreading so rapidly that those who did not make the first trip or two would die at the doors. Others started for the Greene Street exit, but here they ran into the obstacle of the door that opened inward. The crush of people made it impossible for those who had reached it first to open the door. In the struggle, those near the door tried to pull it inward while those in back attempted to shove their way through. At last some of the men in the crowd managed to open the door by brute strength and a few workers escaped.

Those who ran to the Washington Place exit forgot about the outside lock. They piled up in a struggling mass of humanity, tearing each other's clothing as they fought to open the door. Most of them perished as flames overwhelmed them.

Meantime, a bookkeeper had the presence of mind to call the fire department and to notify the tenth floor management offices before she fled headlong from the flames. When Max Blanck got the news his florid face paled. He glanced quickly at his two children, then at the governess. Quickly he ran to the windows and looked down. All he could see was gray smoke pouring upward from the eighth floor windows and a crowd of people already in the street.

There were thirty clerical workers on the tenth floor with Blanck and his partner, Isaac Harris. Blanck's first thought was to ring frantically for the elevator. It came up almost immediately and Blanck allowed ten women to get into the cage. The elevator did not return to the tenth floor after that initial stop because it was too busy rescuing people from the burning eighth floor.

Panic-stricken, Blanck marshalled his two children and the governess into his private office where there was a window from which they might have to jump. Smoke and suffocating heat had now penetrated to the tenth floor and Blanck realized that they could stay there no longer. Then he heard the voice of Harris, excited and high-pitched:

"The roof! Follow me to the roof!"

Blanck, the children, the governess, and the remaining office force groped their way through an already smoke-filled sample room at the rear of the building. All of them managed to escape to the roof where they made their way to a ladder that was slanted against the roof of the building next door. Within minutes all were safe. There is no record that anyone went back to rescue workers trapped in the blaze.

By this time the tiny flame that had been born in a rag bin had turned the entire eighth floor into an inferno. The vicious flames caught those trying to open the locked door at the

Washington Place exit. Amid screams and cries, they were reduced to blackened corpses within minutes. Although a few lucky ones escaped when the Greene Street door was forced open, it promptly closed again from the pressure of the crowd and hungry flames consumed those gathered there. The lone elevator made a couple of trips, saving some of the workers before flames roared down the narrow passage to the elevators and destroyed those who were left.

Some of the women, noticing the hopelessness of the throngs at the exits, and with the heat of the flames at their backs, decided in desperation to jump from the eighth floor windows. The dresses of some caught fire as they stood at the windows, and with screams of anguish they hurled themselves to the street. First one, then two, then a dozen more made the fatal, dizzying leap.

Although the eighth floor was a flaming hell, the crowd below saw only a few wisps of smoke coming from the windows. One of the first to notice the fire from street level was Benjamin Levy, junior member of a wholesale clothing manufacturer in a nearby building.

> I was upstairs in our workroom when one of the employes who happened to be looking out of the window cried that there was a fire around the corner. I rushed downstairs and when I reached the sidewalk the girls were already jumping from the windows. None of them moved after they struck the sidewalk. Several men ran up with a net that they got somewhere, and I seized one side of it to help them hold it.
>
> It was about ten feet square and we managed to catch about fifteen girls. I don't believe we saved more than one or two, however. The fall was so great that they bounced to the sidewalk after striking the net. Bodies were falling all around us, and two or three of the men with me were knocked down. The girls just leaped wildly out of the windows and turned over and over before reaching the sidewalk.
>
> I only saw one man jump. The rest were girls. They stood on the

windowsills, tearing their hair out in handfuls, and then they jumped. One girl held back after all the rest and clung to the window casing until the flames from the windows crept up to her and set her clothes afire. Then she jumped over the net and was killed instantly, like all the rest.

Meanwhile the fire was spreading. The growing flames shot from the eighth floor windows and leaped upward into the open windows on the ninth level. Here they set shirtwaists and scrap material ablaze and roared through the room where some 300 women were employed.

Only a few escaped from the holocaust that swept the ninth floor; there was virtually no way out. The fire leaped across the Greene Street exit within seconds, blocking it. The Washington Place door, as on the eighth floor, was locked.

There was, however, the unsteady fire escape. No one with any choice would have used it. But about twenty girls made it to safety before it was rendered completely useless by fire that licked at the rusty iron and twisted it into warped wreckage.

The only other way to escape was to jump from the windows, as those on the floor below were doing. But this was rendered more difficult because the flames from the eighth level were licking away at the ninth floor openings. Still, many tried it. With clothing and hair ablaze, they leaped by the dozens.

Meantime, fire department units had arrived at the scene of horror. Before they could operate their hoses they had to clear away bodies from the sidewalk. When they did get the hoses in play, women jumping from the building landed on them and had to be removed so that the hoses could operate. The fireman's net collapsed as bodies rained down on it. Then, to compound the disaster, the firemen discovered that their ladders reached only to the sixth floor and the hoses shot a stream of water only as high as the seventh!

Samuel Levine, a machine operator on the ninth floor who finally escaped the fire, told a story of courage and imagination:

> I was at work on the ninth floor when the fire began. The girls on the floor dropped everything and rushed wildly around, some in the direction of the windows and others toward the elevator doors. There were flames all around in no time. Three girls came rushing by me. Their clothes were on fire. I grabbed the fire pails and tried to pour water on them, but they did not stop. They ran screaming toward the windows. I knew there was no hope there, so I stayed where I was, hoping the elevator would rescue me. When the elevator didn't come up I smashed open the elevator doors. I guess I must have done it with my hands. I reached out and grabbed the cables, wrapped my legs around them, and started to slide down. I can remember getting to the sixth floor. While on my way down, as slow as I could let myself drop, the bodies of six girls went falling past me. One of them struck me and I fell to the top of the elevator. I fell on the body of a dead girl. My back hit the beam that runs across the top of the car.

Levine was rescued from the elevator shaft after the fire was out and firemen were searching for bodies. Rescued with him was Cecilia Walker, a twenty-year-old girl who had also descended by way of the cable. Another man, Hyman Meshel, was found in the elevator shaft below the floor of the elevator, standing in water up to his neck whimpering like a trapped animal.

Those who jumped down the elevator shaft made it impossible for the elevators to continue operating. At least thirty bodies eventually clogged the shaft.

The most frustrating feature of the fire was the helplessness of the firefighters. Although they had answered the alarm promptly, by the time they arrived the three top floors of the building were ablaze. Without adequate ladders and hoses, they could do nothing but try to catch those who jumped. Seeing the futility of the firemen, men in the gathering crowd across the street tried to force their way through police lines to help the victims, but they were held back. They could have done little in any case.

It was as if the heavens were raining rejected bodies on the earth. They came singly and in batches. One group of five women jumped as they clung to each other. All were killed. Another hung from a windowsill until the flames licked at her fingers, then dropped to her death. On the ninth floor a man and woman appeared at a window. He embraced and kissed her, then hurled her to the street. He jumped after her and both died.

Although the crowd kept yelling, "Don't jump!" the bodies poured down. One policeman on the sidewalk stared at a charred corpse crumpled in the street. It was headless.

"I saw the *Slocum* disaster," he said. "This is worse."

"Is it a man or a woman?" asked a reporter.

"It's human, that's all you can tell."

Frank Fingerman, who worked in a nearby building, turned in a fire alarm from a Broadway box when he first heard the cries from the building.

"As I ran to the fire box," he told a reporter, "I saw a boy and a girl standing together at a Greene Street window. She seemed to be trying to jump, but he was holding her. They were still there when I came back from the fire box. As smoke began to come out of the window above them the boy let the girl go, and she jumped. He followed her before she struck the ground. Four more came out of the same window immediately."

Pauline Grossman, an 18-year-old worker, told a story of heroism during the fire. She said that three male employees had formed a human chain across a narrow alley between buildings. Several persons passed across the men's bodies, escaping from the burning factory and entering the window of the building opposite.

"As the people crossing upon the human bridge crowded

more and more over the men's bodies," she said, "the weight upon the body of the center man became too great and his back was broken. He fell to the passageway below and the other two men lost their holds upon the windowsills and fell. Persons who were crossing the human bridge dropped with them to the alley."

In a building across the narrow courtyard from the Asch Building, twenty New York University students were attending a lecture. One student, Frederick Newman, recalled: "We were in the library of the building on the top floor when we noticed a gust of smoke coming from the building across the courtyard. Sparks drifted in at the open library window and as we jumped from our seats we saw the girl workers crowding at the windows. We saw a man leap and then the girls began to follow him."

The students quickly found two ladders and positioned them so that one end rested on the windowsills of the burning building and the other end on the library window. Several young women made it across the perilous ladder while others, with hair and clothing afire, leaped into the courtyard to their deaths.

O. B. Smith, another student who was late arriving for the lecture, was stopped by a police line at Greene Street. "Across the street I could see the bodies of five women in a pile," he said. "As I looked I saw an arm raise, and I knew that one of the women was alive. I called out to a policeman standing near. His only answer was, 'Get back there and mind your own business.' I pointed out the woman to him and said something ought to be done, as water from the firemen's hoses was pouring down on her. Perhaps he didn't understand me, for nothing was done."

This may have been the same woman that an officer, who was later checking bodies as they were shipped to the morgue, found alive among a heap of dead. He pulled her free of the corpses and placed her on the sidewalk, but she died before an ambulance could take her to a hospital.

When the fire was over the Asch Building looked much the same as it had before—from the outside. It was a "fireproof building" in the sense that it did not burn to the ground or crumble from the flames. But the top three floors were gutted, and in the inferno one hundred forty-five perished.

Fire Chief Croker was probably the most bitter man in New York City. He had predicted that a loft fire would occur and that it would be a disastrous one, but no one had paid attention. As he and others inspected the ruins, he said, "Look around here. Nowhere will you find a decent fire escape. They say they don't look sightly, but I have tried to force their installation and only last week a manufacturers' association met in Wall Street to oppose my fire protection plan which included a sprinkler system as well as additional fire escapes. This is just the calamity I've predicted. I have been advocating for a long time that more fire escapes be put on buildings like this. The large loss of life is due to this neglect."

The horrified reaction from the American public created a scramble among officials as they tried to place the blame elsewhere. Various agencies of city and state government were criticized—the New York Department of Buildings, the State Labor Department, the Water Supply Department, the Police Department, the Fire Department. All of them denied responsibility.

When the Waistmakers Union organized a mass funeral for the victims, ten thousand mourners attended. New York's East Side, from which most of the dead had come, seethed with anger. On April 5 more than eighty thousand protesting men and women marched up Fifth Avenue, following an empty hearse pulled by six horses draped in black.

One beneficial effect of the tragedy was that the New

York State Legislature appointed a Factory Investigation Commission to examine the situation in factories throughout the state. The result was legislation that improved factory conditions everywhere. In the city of New York, more than thirty ordinances were enacted enforcing stringent fire precautions in factories.

The owners of the Triangle Shirtwaist Company, as so often happens, endured little punishment for their laxity. On April 11, both Max Blanck and Isaac Harris were indicted for first and second degree manslaughter. Their trial did not come up until December 28, by which time some of the horror had dissipated. When questioned about the practice of keeping the Washington Place exit locked from the outside, both said they were unaware that the door was locked. The judge accepted this statement and told the jury that they could not be found guilty unless it was proved that they knew the door was locked.

It couldn't be proved, and on this technicality Blanck and Harris were acquitted.

7

The Halifax Horror (1917)

THE French munitions ship *Mont Blanc* moved slowly through Halifax Harbor and cautiously poked her bow into the six-mile-long Narrows that separated the outer harbor from Bedford Basin where she would be temporarily docked. In her hold were 5,000 tons of trinitrotoluene—TNT—along with such chemicals as picric acid and benzol. The *Mont Blanc,* just arrived from New York, was scheduled to join a convoy at Halifax, Nova Scotia, en route to Europe where she would deliver her dangerous cargo to Allied troops fighting Germany in World War I. The 3,121-ton munitions freighter, under Captain Lamodec, was being guided through the mile-wide Narrows by Frank Mackie, an experienced pilot from Halifax.

At about the same time the *Mont Blanc* entered the Narrows from the south, the 5,043-ton Norwegian freighter *Imo,* headed for Belgium with a load of grain, entered from the north. There was plenty of room for two ships to pass, but miscalculations had occurred in the past and caution was necessary to avoid minor collisions and scrapings. However, it was broad daylight now and visibility was excellent, and the two ships were expected to pass without incident.

The fateful morning of Thursday, December 6, 1917, began no differently from any other. At nine o'clock children were already seated at their schoolroom desks; mothers had turned their attention to household duties; employees of factories, warehouses, and offices were on the job; doctors were attending wounded soldiers at the military hospital; politicians were busy at the City Hall and Provincial Parliament Building; the cotton mill and sugar refinery both hummed with activity; longshoremen loaded the ugly materials of war on waiting ships; Canadian Pacific Railroad trains rumbled into and out of the depot; trucks and wagons clattered noisily through the narrow streets. It was the start of a typical wartime day.

The Great War had, in a sense, been kind to Halifax. It had catapulted the city from an unimportant port to one of the greatest gates of ocean traffic on the North American continent. Her harbor was constantly filled with ships carrying munitions and other supplies to the Canadian and American forces in Europe. Proud of its role in the big war, Halifax was also content to be a haven from the bitter struggle that raged so far away. Canadian coast artillery was on alert both day and night to protect the city. After dark, great searchlights spanned the sea, and the busy harbor was protected by huge nets to keep out German U-boats. Although many of its young men had been sent to Europe to do battle, there seemed little chance that the city would suffer directly.

But in the Narrows the two ships that were to bring almost unbelievable catastrophe to the city approached each other like wary boxers seeking an opening. Slowly they closed the gap, each vessel at first keeping to the right or starboard side of the channel. If they had pursued this course, there would have been no problem. However, when each ship used its whistle to signal its intention to the other, confusion resulted. Survivors on the *Imo* insisted that the *Mont Blanc* cut across the channel in front of the Norwegian ship. Other witnesses said the *Imo* changed course suddenly and headed directly for the *Mont Blanc.* In any case, a collision occurred.

The *Mont Blanc*'s Captain Lamodec, realizing the inevitability of the collision, tried to turn his ship so that the bow would take the brunt of the blow. He knew the kind of cargo he was carrying in the hold, and he was aware that the five thousand tons of TNT could explode from a severe jar. He succeeded in his last-second maneuver and the *Imo* struck the *Mont Blanc,* not with great violence but hard enough to rip open the munition ship's forward hull.

The TNT did not explode. But the impact ignited twenty barrels of the benzol stored in the forward part of the ship. Immediately the blazing benzol began running over the deck and seeping into the hold. Captain Lamodec knew what would happen. When the burning fuel reached the TNT, there would be a frightful explosion and the ship would be torn apart.

Captain Lamodec looked down from the bridge and saw that his competent French crew had already rolled hoses and were attempting to control the spreading blaze. But he also saw that the situation was hopeless. In minutes the blazing fuel would reach the TNT. The only course of action was to flee the ship and hope to get far enough away to escape the terrible force of the blast.

"Over the side!" he shouted. "Abandon ship!"

Meantime, the *Imo*'s Captain Fron, thinking his ship's plates had buckled, ordered his engines reversed. Within a couple of minutes the *Imo* had backed off and successfully beached herself. Sitting at what he felt was a safe distance from the wreck, Captain Fron watched in amazement as the crew of the *Mont Blanc* quickly lowered lifeboats and rowed frantically for shore. He did not know what the *Mont Blanc* carried in her hold, and seeing the entire French crew abandon ship after a relatively minor collision made no sense to him.

There was another ship close by whose crew witnessed the collision. She was the British cruiser H.M.S. *Highflyer,* slated to lead the convoy of which the *Mont Blanc* was to be a part. Her commander knew precisely what was in the hold of the *Mont Blanc* and was well aware of the terrible damage that would result if the fire on her deck reached the hold. Moreover, he thought that the French crew had abandoned ship too hastily, and there was still a possibility that the fire could be extinguished before it reached the TNT.

"I'm calling for volunteers to put out that fire!" he shouted.

Immediately twenty men stepped forward. The commander and his men went over the side at once, rowing a lifeboat toward the burning ship. Climbing aboard, they grabbed hoses and tried desperately to put out the blazing benzol. It was a display of courage that stands high in the annals of naval history, but the attempt was doomed to failure.

Meantime, the crew of the *Mont Blanc* was rowing in two boats toward the north shore of the channel, where dockworkers regarded them with only tepid interest. From Pier 8, they could see nothing wrong with the *Mont Blanc*. The flaming benzol streaming across the deck was not visible from the shoreline, and why the Frenchmen were so frantically rowing the lifeboats was a puzzle.

The lead boat hit shallow water with a crunch and the crew stumbled ashore. They shouted warnings to the townspeople in French, but few of their listeners understood. One sailor with a smattering of English pointed to the *Mont Blanc* and said, "Powdair! Powdair!"

When the second boat landed the people learned what the danger was. In this boat was the Halifax pilot, Mackie.

"The *Mont Blanc*'s afire! She's a munitioner!" he cried. "Run for your lives!"

The crewmen raced through the streets in a headlong dash that startled bystanders, and their dreadful news spread quickly. Some, realizing the extent of the danger, began to flee with the Frenchmen. Others, not knowing what to do or underestimating the danger, stayed where they were. Even some of the longshoremen stood on Pier 8 trying to see the flames on the deck of the *Mont Blanc*.

Then, just seventeen minutes after the collision, the *Mont Blanc* exploded with one of the greatest blasts in history up to that time. In the fiery upheaval the volunteers from the H.M.S. *Highflyer* perished instantly.

The noise of the explosion was so loud that survivors said later that they had never heard anything even remotely like it. A huge, black cloud of smoke six hundred feet high billowed into the sky and blotted out the sun. The *Mont Blanc* simply disappeared in an inferno of smoke and flame. The beached *Imo* was plucked out of the shallow water and hurled ashore like a toy.

The tragedy that overwhelmed Halifax was due not only to the explosion and the resulting fire but also to the topography of the region. Slopes, rising on either side of the Narrows, contained the fierce pressure of the explosion, and it was that part of the city lying in this trough that took the brunt of the blast. On the south shore lay a heavily populated section called Richmond, a clutter of wooden homes where the laboring class lived. This section was virtually wiped out.

As the mushrooming smoke cloud began to settle over Halifax like a great shroud, an upheaval of water from the channel drove surging waves for two blocks up the sloping sides of the city. Some people were drowned immediately in the rushing water, and others were carried back to the channel by the receding waters, and swept to their deaths.

Destruction and death were everywhere. The walls of the Richmond School caved in and killed two hundred children who were just starting morning classes. Flimsy houses shuddered and fell into ruins, burying their occupants. Public buildings were flattened. Huge trees were catapulted into the air. People caught in the violent blast were lifted into the air and smashed against walls and telephone poles. Freight cars were blown through the air for a distance of two miles. Windows in houses miles from the blast were shattered.

Great metal plates from the *Mont Blanc* were torn from the hull and hurled a mile and a half from the waterfront, falling on houses and people. A piece of her anchor, weighing half a ton, sailed through the air and crashed to the earth three miles away. A five-pound piece of jagged metal struck the baggage car of a train two miles from Halifax and killed a brakeman instantly. Other portions of the ship—chains, rivets, pieces of the deck and superstructure—rained over the city. Telephone poles flew about like straws, and even those a mile and a half away were either uprooted or broken off at ground level.

People lingering on the waterfront had little chance. Most were incinerated. Many killed by the blast were stripped of their clothing. A sailor was blown one thousand feet from the harbor, landing with nothing on his body but his boots. Worshippers in nearby St. Joseph's Church were killed as the walls fell in.

The force of the giant blast, acting like an immense cyclone, spread havoc over the waterfront. The *Stella Maris,* a salvage ship near the *Mont Blanc,* was torn to shreds and driven ashore, its captain and crew dead. The blast tipped over or incinerated small vessels; sailors fell from the riggings of nearby ships and caught fire before they reached the water; ships at their moorings were torn free or engulfed in flames;

one naval ship's turret was splattered with the bodies of eight sailors slammed against it; a great rock was lifted from the bottom of the Narrows and tossed like a pebble against a ship, killing sixty-four sailors. Not even the French seamen who had deserted the *Mont Blanc* were all saved; many of them died in their tracks as hot fragments of steel from the ship struck and killed them.

In a shed on the waterfront, Duncan Grey was inspecting shells. Hearing the excitement on shore as the Frenchmen landed in their lifeboats, he rushed out of the building. Somehow he survived the explosion, which he described:

> A few seconds after the roar of the explosion a gust of wind swept through the shed and down came pillars, boards and beams. I was in the open and the sight that met my eyes was the worst I hope ever to see in this world.
>
> I have been in the trenches of France. I have gone "over the top." Friends and comrades have been shot in my presence. I have seen scores of dead men lying on the battlefield, but the sight that greeted me now was worse and far more pathetic. I saw people lying around under timbers, stones and other debris. Some were battered beyond recognition and others were groaning in their last agonies.
>
> Rushing here and there, I struggled to assist them, and as near as I can remember, I pulled out twenty-two men from under wreckage. As I was right in the affected district, I witnessed the full horror of the situation. Partly blinded by the smoke from burning dwellings, I groped around, assisting some of the poor mothers and little ones who were running about screaming and searching vainly for lost ones, in many instances never to be seen again. I struggled on, coming across more and more bodies of dead men, dead women, dead children. Death was everywhere.
>
> Flames were sweeping a wide pathway for themselves. Doomed structures were belching forth great volumes of smoke from doors and windows. The district was a living hell.
>
> Half strangled by the smoke, I kept pulling out bodies from under beams and fallen chimneys and wreckage. Some of the

bodies were without clothing. Many were so mutilated that it was difficult to realize that they were human. Some men were virtually demented. Thinking only of their wives and children, they dashed about in the burning debris, hazarding their lives with the single thought of finding their own.

I shall never forget how I felt in that hour. I saw little kiddies running along, some with blood streaming from them. All were crying for their parents, while mothers and fathers raced about in frenzy. I have never seen anything so pathetic, even on the battlefield.

The streets of Halifax indeed were in utter confusion. Automobiles had been thrown against buildings, street cars tipped over. The railroad station crumbled, burying hundreds of waiting passengers under heavy masonry. Women and men working at their looms in the cotton mill were crushed when the building crumpled and heavy machinery fell through the floors. The cotton burst into flames that leaped to a row of nearby tenements and burned them out.

The heat from the explosion ignited fires everywhere. Wooden homes were natural targets, and many families burned to death instantly. The sugar refinery went up in lurid flames. So did a dozen or more giant warehouses. Scattered fires merged to form a single holocaust that ate away at the crumbled ruins caused by the blast.

The entire area from Pier 8, near where the *Mont Blanc* blew up, to the North Street bridge—a distance of more than two miles—was burned out. People stumbled through the fiery ruins to become pillars of flame; people with broken legs and other injuries lay in the streets until fire swept over them. Animals—dogs, cats, horses—became torches. A great stench rose over the destruction, the odor of bodies burning.

All over the city the after-blast of the explosion rolled on relentlessly, leveling almost everything in its path. Building after building toppled, walls crashed, windows shattered, showering people in the streets with jagged pieces of glass.

Refugees from Richmond district and the harbor area, who had somehow managed to escape with their lives, fought their way through the streets in an effort to get as far away from the roaring flames as possible. Many were stained with blood; some had lost their hands or arms; others, blinded, were assisted by loved ones. One woman, leading a child, held the headless body of her baby in her arms.

An almost unbelievable act of heroism saved Halifax from even greater destruction. The British freighter *Picton,* heavily laden with munitions, was docked close to the *Mont Blanc* when the explosion took place. The great blast ripped and tore at the ship, almost lifting it out of the water and killing the skipper and most of the crew instantly. The few sailors still alive abandoned ship and fled ashore.

On the deck of the *Picton* flames began to spread. It would have been only minutes before they reached the hold and set off a blast as terrible as the one that had just occurred. Except for one lone man—Captain J. W. Harrison, marine superintendent of the Furness-Withy Line at Halifax.

Harrison knew the cargo that the *Picton* carried. From the shore he saw flames dancing on its deck and, without regard for his own life, went into action. He sent a messenger into the streets to warn everyone to run as far from the dock as possible. Then he raced to the ship, boarded her and cut the hawsers to set her adrift. With the ship drifting into the channel, Harrison went below and opened the seacocks in the hull, hoping to prevent an explosion by letting in water. Then he returned to the deck, attached a reel of hose and played the stream on the blaze until it was out. When he was later asked to describe his experience, Harrison declined, saying only that he had done his duty as he saw it.

As in all great tragedies, rumors spread through the stricken city almost as fast as the flames that engulfed it. One widely believed story was that a German fleet was bombarding the city. Some said that the explosions and fire were the result of an air raid, and there were people who claimed they actually had seen a German aircraft in the sky. One man insisted that a German shell had streaked past him, bursting some yards away. Such unfounded rumors naturally terrified the townspeople. Thousands fled the city into open country; five thousand jammed Halifax Common.

The rest of the world would not have heard about the Halifax tragedy for some time had it not been for Vincent Coleman, a telegraph operator at the Richmond station. He saw the collision of the *Mont Blanc* and *Imo* and immediately transmitted a message to the world: *A munition ship is on fire and is making for Pier 8. Goodbye.* Later Coleman's body was found at his post. Further communication was cut off by the terrible detonation. But in the small town of Amherst, Nova Scotia, a telegrapher sent a message stating that there had been "some kind of explosion" in the harbor of Halifax.

A man who identified himself only as Lieutenant-Colonel Good, of Halifax, told a reporter:

> All that could be seen for miles in circumference were burning buildings, great mounds of iron and brick in the streets, dead bodies strewn along the sidewalks, the living with the dead. Many had fractured skulls or broken limbs and were dying unattended in the streets.
>
> I set out to aid the injured, and for a moment stopped before a wooden frame house that was in flames. An old man stood helpless, crying out that his wife was in the burning house somewhere. With two others I managed to enter the house and found what appeared to be the half-burned body of an old woman. We brought it into the open and beckoned to the aged man. For several seconds he stared as though bewildered at the smoking pile. Then he said, "Well, that's my wife." There were hundreds, many hundreds, of similar incidents.

William Bartou, a traveler, was at breakfast in the Halifax Hotel when the blast occurred.

> In ten seconds it was all over. A low rumbling, an earthquake

Complete devastation, with only portions of buildings standing, testifies to the widespread damage caused by Chicago's great fire of 1871.
(Courtesy Chicago Historical Society)

Fire engines (vintage 1903) in the alley behind the Iroquois Theater attempt to get at the origin of the fire on stage. The street exit was jammed with the dead and dying. *(Courtesy Chicago Historical Society)*

Hundreds of people—mostly women and children—perished as fire ravished the excursion boat *General Slocum*, the gutted remains of which are shown here. *(United Press Photo)*

Clouds of black smoke hover over San Francisco as the earthquake-sparked fire destroys 490 city blocks and damages 32 more.

(Courtesy California Historical Society, San Francisco)

Fire-blackened debris and charred walls are all that are left of the eighth floor of the Triangle Shirtwaist Company where a tragic fire killed 145.

(United Press International Photo)

The complete devastation at the waterfront in Halifax graphically portrays the power of the explosion and fire that killed 1,600 and injured 8,000. *(United Press International Photo)*

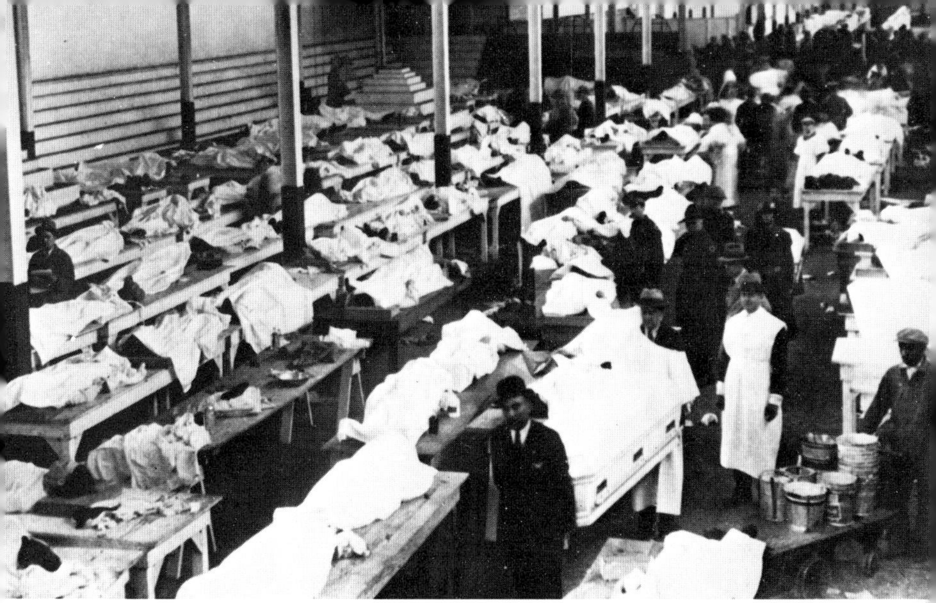

The bodies of the convicts burned to death in the Ohio State Prison fire in April 1930 lie in the emergency morgue set up on the fairgrounds.
(United Press International Photo)

The silver envelope of the dirigible *Hindenburg* bursts into flames at the mooring mast at Lakehurst, New Jersey. Within minutes the blackened skeleton of the ship was on the ground. *(United Press International Photo)*

Confusion reigns as rescuers try to help victims of Boston's Cocoanut Grove Night Club fire in which 493 merrymakers perished.
(United Press Photo)

An oval of charred remains, including the fire-twisted bars of the animal cage, bears testimony to the terrible Ringling Brothers fire that destroyed the Big Top and sent 168 to their deaths. *(United Press International Photo)*

Fire gutted the 15-story "fireproof" Winecoff Hotel in Atlanta, trapping 285 guests and killing 119 of them. *(Acme Photo)*

Vast destruction was caused by two explosions on ships loading ammonium nitrate at Texas City's busy wharf. The explosions and subsequent fires virtually destroyed the city, killing 561 and injuring 3,500. *(United Press International Photo)*

Firemen run up ladders and strive to rescue children trapped in the heart-breaking fire at Our Lady of the Angels Parochial School. Ninety-

> shock, with everything vibrating, then an incredible noise, followed by the fall of plaster and the smashing of glass. In such moments the human mind does not hesitate. A cry went up: "A German bomb!" A rush for the door, headlong down the hallway with falling pictures, glass and plaster, to the swing doors of a few moments before now ripped from their hinges, through great projecting triangular pieces of glass to the street. Here I found myself with a burden. How she had come into my arms I do not know, yet here she was, hysterically shrieking, "Oh, my poor sister, my poor sister!"
>
> Outside, overhead, a giant smoke cloud was moving northward. The danger seemed over. I crossed the road, placed my feminine burden on a doorstep, and returned to the hotel. My aid was possibly needed more there. I made my way upstairs to the rooms of two friends. The rooms were vacant but normal, even the glass unbroken—and few panes of glass remained unbroken in the area. Once more I was in the street, meeting my companions on the threshold. They, like me, were unscathed.
>
> Our plans were quickly made. We were off to the immediate vicinity of the disaster, for, among many theories, we accepted as most plausible the blowing up of a munition ship. Toward Citadel Hill we wended our way and the farther we went the more horrid it all became. The wounded were everywhere, but since most of these unfortunates could hobble or walk, we kept onward. Our hurry-scurry led us to the armory. Here the khaki-clad men were, despite wounds they suffered, on the march toward the more devastated areas. The order had gone forth, "Commandeer all vehicles, autos and horses." A cordon was drawn across the streets and passengers were forced to alight and resume their journeys on foot. There was grim work ahead.

To the credit of citizens like these, the "grim work" began immediately. Even while the bulbous cloud of smoke still hung low over the city, rescue efforts got underway. Troops moved out of their barracks so that women and children would have a place to stay. Five hundred tents were pitched on the Halifax Common. Firemen, soldiers and volunteers moved into the burning area and began extinguishing the fires with water from the channel. The wounded were taken to a central location where doctors worked over them. The dead were piled like cordwood and carried away in wagons.

One group of soldiers found a still-living young girl pinned beneath the timbers of a collapsed house. They worked for an hour to release her, but as they lifted the last of the wreckage, she died. A six-year-old girl, found wandering about helplessly, told an amazing story of escape. She had been blown through the roof of a house, rolled from the roof to the ground, and suffered only a few scratches. Another group of rescue workers found the body of a ten-year-old girl still clutching her school book.

Eamond P. Barry, a postal clerk, described his experience as a rescue worker in the hard-hit Richmond section: "People in our area died like flies. We found some with noses shot off, eyes put out, faces slashed from flying glass, limbs torn and distorted. On one occasion, while we were working around a wrecked building, we could see a little baby fifty feet or more underneath a burning mass, crying for help. We could not get within thirty feet of the child and had to watch while it burned to death."

On the outskirts of Halifax, some distance from the blast, was Mount Vincent's Academy, a convent that was hastily converted into a hospital. The injured were cared for by the sisters and girl students who attended the academy. Eleanor Tapley, a student, recalled:

> We first received word of the disaster from an engine which came hurrying up from the city, rocking from side to side under its terrific speed. The engineer, the only man aboard, cried out, "Give me anything you have. Food, blankets, bandages, or anything. The whole city is wrecked and for the mercy of God be quick!"
>
> We girls immediately rushed to get anything that we could lay our hands on. Sweaters, coats, and other clothing were torn in strips for bandages. Everything was piled into the locomotive which then tore away again at top speed for the scene of the disaster.

> No one at the convent was killed, but some of the sisters were cut by flying glass, which in our section of the city was the most damage. Every window in the academy was broken and some pillars in the chapel fell.

All day long the rescue workers labored at their task of helping the living and carrying away the dead. When night fell, only torches and lanterns lighted their way. Doctors performed operations under the dim light of oil lamps. Every building standing was converted into either a hospital or a morgue.

When daylight came again there were still hundreds buried under fallen debris—many dead, some still living.

The heart-breaking work went on. Soldiers and sailors patrolled the street to discourage looting. All day long weary rescuers searched for new victims.

Then, as night closed in again, a second disaster struck the battered city. As if to finish the gruesome job the explosion had started, a raging blizzard—said to be one of the worst ever to hit the city—roared down on Halifax. Icy winds numbed the rescuers. Blinding snow impeded progress. The city was blanketed in impenetrable darkness. It was impossible even to keep lamps lit in the gale that swept over the ruined city. In the temporary hospitals, snow swept in through broken windows and the cold wind whistled through the rooms. Many who otherwise might have recovered died that night from exposure or contracted pneumonia and died within a day or two.

Meantime, the outside world finally began to realize the extent of the Halifax horror and sent help. The Canadian Government sent rescue workers and supplies. Special trains arrived from New York and other cities.

Slowly Halifax recovered from its ordeal. The official statistics were horrendous: sixteen hundred killed, eight thousand injured, two thousand missing, three thousand dwellings destroyed, 20,000 destitute; loss of property was estimated at $30,000,000. When the rescue work was completed, there were five hundred people unaccounted for—the number probably blown to bits in the explosion. More than half of the city had been destroyed.

Several investigations of the catastrophe were undertaken by agencies of the Canadian Government, and efforts were made to place the blame for the explosion. Since the benzol on the *Mont Blanc* had caught fire and ignited the TNT, a serious question arose as to whether such inflammable material should be carried on the same vessel that carries munitions. Some naturally asked why it was even necessary to bring such cargoes into a harbor like that of Halifax; they accused the government of ignoring the safety of its citizens by subjecting "a priceless storehouse of materials of war to the confusion of whistles in a crowded harbor."

But when the investigations had been completed, one court found the *Mont Blanc* at fault, another the *Imo*. Later the highest court in the British Empire took the easy way out. It blamed both ships equally.

The *Mont Blanc,* of course, was wrecked beyond salvage. But the *Imo* was repaired and went to sea again. On December 6, 1921, she struck a hidden reef on a voyage to the Falkland Islands and sank—exactly five years to the day after the catastrophic collision in Halifax Harbor.

8

Ohio's Prison Holocaust (1930)

SUPPER was over. Most of the gray-clad prisoners in the west cell block of Ohio State Penitentiary at Columbus had been marched back to their box-like cells and locked in for the night. To kill the hours before lights would be turned out, those in the upper tiers read books from the prison library or settled down to games of cards or checkers. Those prisoners occupying the lower tiers had not yet been locked in. They moved in a long sluggish line under the alert eyes of the prison guards, contemplating the dread monotony of another evening in their bare cells.

It was a routine maneuver, this locking up of some 800 prisoners in the six-tiered cell block of the sprawling penitentiary. But at that moment a flashing tongue of flame was insidiously eating away at the northwest corner of the roof. It was the beginning of a raging holocaust that was ultimately to take the lives of 320 trapped men and injure 230 others—in one of the most appalling fire tragedies in prison history.

It was April 21, 1930. The Great Depression had already

delivered a stunning blow to the American people. Men who had worked steadily for years suddenly found themselves without jobs and with money running low. Young people fresh out of high school and college faced the grim specter of idleness. Under these conditions crime was becoming rampant and already most prisons were overcrowded. Big, gray-walled Ohio State Penitentiary—one of the largest prisons in the country but also one of the most badly crowded—housed forty-three hundred inmates in accommodations built for only fifteen hundred.

Warnings about the dangers of such overcrowding had been sounded repeatedly for years. Veteran Warden Preston Thomas had urged twelve successive Ohio State legislatures to take action, but mostly the legislators had shown only bland indifference. In 1929 the National Society of Penal Information joined the chorus of protests, singling out the prison as one of the country's "horrible examples." Its report stated: "The need of another institution in the Ohio penal system has been apparent for many years, but the state is only now taking steps to alleviate the condition. Even with the completion of the present building program, it will be able to care for its present population only under conditions that fall far below accepted standards for housing prisoners."

The "present building program" consisted of a pitifully inadequate addition to the west cell block. A maze of scaffolding had been erected outside the older building, and it was here that the uncontrollable fire that swept the prison on that grim April evening started. Fanned by a brisk breeze, the feeble first flame became a ravenous monster. It climbed the scaffolding with the agility of an acrobat and leaped across the roof of the west cell block. Here it found plenty to feed upon. The roof of the prison was made of wood covered with tar paper, a startlingly incendiary package considering that officials had always insisted the prison was fireproof.

Convict Barry Sholkey spotted the flames first. Already caged in his cell, he saw the saffron flames through a window as they roared up the scaffolding. Wildly, he rattled the door of his cell.

"Fire!" he screamed. "The prison's on fire!"

A guard walking by stared at him dubiously.

"Shut up, Sholkey," he said simply.

Sholkey's beady eyes held a fanatic light. He moistened his lips nervously.

"I tell you the prison's on fire!"

The guard grinned.

"Always the clown, ain't you?" he said.

It was true. Sholkey was given to kidding and playing practical jokes. Nobody believed him anymore, and prisoners who had heard the exchange roared with mirth.

"Pipe down, you guys!" said the guard.

Minutes later they were taking Sholkey seriously. In this incredibly short time the fire had leaped to the roof, had eaten its way through the tar paper and timber, and had begun to shower sparks on the occupants of the upper tier of cells.

The place became a bedlam. The flames were now visible as they consumed the roof. Bits of burned timber filtered down, and soon bedding and mattresses were blazing. The men, trapped in their cages, screamed and rattled the doors. But their cries fell on unheeding ears. The guards, following regulations to the letter, refused to unlock the cells.

In the lower tiers, many of the prisoners still had not been locked in. With smoke beginning to billow through the cell block and terrified cries from above ringing in their ears, the men refused to enter their cells. The guards, determined to follow the decree that all prisoners had to be locked in by six o'clock, went about their duty. The prison was fireproof, wasn't it? It was only a small blaze and would be put out within a few minutes.

As guards shoved the prisoners toward the yawning doors of the iron-barred cages, a sullen roar went up from the frightened men. Suddenly all was confusion. Guards were seized and thrown to the floor. Pitched battles took place in the huge cell block as the guards tried to stem the rebellion.

At first the prisoners sought only to drive the guards back so that they would not be forced into their cells. The convicts had no weapons, but larger numbers and flailing fists quickly turned the tide. Step by step the guards gave way, backing slowly toward the door leading from the burning building to the yard. There they made a last desperate effort to stave off the mutinous prisoners, but they were either tossed bodily out of the way or pummeled to the floor.

The smoke was thickening now and both convicts and guards choked and strangled on it as they fought. The density of the smoke drove the convicts to greater effort, and within minutes they had opened the door to the yard and were pouring out into the fresh, cool evening air.

But on the other tiers, many of the convicts were still locked in. Two black prisoners, John Sherman and Charley Simms, thought of these trapped men—and became heroes. Having been separated from the body of fighting convicts, they headed for a second door leading to the yard. Here they found a guard steadfastly attempting to keep them from leaving the burning building.

Sherman and Simms moved slowly through the swirling smoke and, under its cover, grabbed the guard and bore him to the concrete floor. Sherman ripped his keys from him. Then, ignoring the door to safety, the two men raced back into the dense smoke and began unlocking the cells of those who had been trapped.

They had rescued 68 men and herded them to safety before they finally stumbled outside themselves, almost overcome by smoke.

But in the uppermost tier, stark tragedy was unfolding. There were 168 men in this tier, all locked behind bars. The roof overhead was crackling and spitting flames, turning each cell into a raging oven. The men, choking in the heavy smoke and dodging the rain of fire, yelled frantically to be let out.

Thomas Watkinson, head guard of the upper tier cells, made no move to open the doors. He stood in the mushrooming smoke and gazed at the men as if in a trance. Another guard rushed up and urged him to let the men out. Watkinson shook his head stubbornly.

"No. I have orders not to open the cells," he said.

Another guard came up with the same plea, but Watkinson refused to relinquish the keys.

The men had begun to die. One, suffocating in the heavy smoke, coughed feebly and sank in the corner of his cell. Another, struck by a blazing timber, turned into a human torch. Some, to reach fresh air, climbed the bars of their cells like monkeys but fell back in agony as the hot metal scorched their hands. Those still alive screamed in horror, their eyes streaming tears, and shook the doors of their cells like men gone mad.

Through the thick clouds swirling around them, the two guards—Thomas Little and George Baldwin—approached Watkinson once more for the keys. As Watkinson backed against the hot bars of one of the cells, the two men rushed him. Writhing and struggling, Watkinson tried to get away. Finally one of the guards was able to wrest the keys from him.

It was too late. To their horror the guards found that the keys no longer opened the doors—the intense heat had welded the locks and the keys could not be inserted!

"Bring crowbars!" somebody shouted. "Open the doors!"

In the cells the men continued to die, agony on their faces as the flames engulfed them. Then the ultimate tragedy occurred. The flaming roof that had been showering the inmates with death-dealing sparks collapsed completely. It came down with a hideous roar, pinning them in fiery debris. Little and Baldwin watched the doomed men helplessly.

Not one prisoner of the 168 in the upper tier escaped the holocaust.

Another guard later described the upper tier tragedy in these horror-filled words: "I saw their faces, wreathed in smoke that poured from their cells. With others I tried to get them out, but we could not move the bars. Soon flames broke into the cellrooms and the convicts dropped before our eyes. They were literally burned alive."

By now the fire was spreading to other tiers where prisoners and guards had begun working side by side to rescue the entrapped men. In the confusion, keys were lost and others failed to fit heated locks. Sledges and crowbars were brought in, and prisoners pounded desperately at the locks. One over-zealous guard, seeing a prisoner trying to open a cell door with a chisel, drew his pistol and shot him dead. But for the most part everyone worked together—until the heat and smoke became so intense that both rescuers and rescued began dropping to the floor and dying of asphyxiation.

There were many cases of heroism among these outlaws of society. A lifer called "Wild Bill" Croninger, a notorious gunman who had escaped from the building, heard the cries of doomed men above the savage roar of the flames and raced back into the fiery furnace. He found a man lying outside his opened cell door, gasping in the thick smoke, and carried him outside. Then, heedless of his personal peril, he rushed into the building again.

The intense heat almost took his breath away. But he hesitated only an instant and then ran forward through the impenetrable haze until he collided with another convict stumbling about aimlessly. Hoisting him on his back, Croninger carried him out of the building.

A third time he went back. This time a panic-stricken

prisoner tried to fight him off, thinking in the confusion that he was a guard trying to prevent his escape.

Croninger hit him solidly on the jaw, toted him, unconscious, from the burning cell block.

In all, Croninger dragged or carried twelve men from the raging inferno. When he went back in for the thirteenth time the heat cauterized his lungs. He fell to the floor, crawled painfully to a corner, and died.

Narcissa Gaeta, who had been serving time for armed robbery, was another hero. This big brute of a man had carried fifty prisoners to safety when on his last attempt, smoke blinded him. He staggered around trying to find the exit, and at last headed for the place that had been his home for seven years—his cell. Groping his way, he stumbled inside, his eyes streaming tears from the smoke. Then he sat down in the corner; oblivious to the flames leaping around him. Later Dr. George W. Keil, chief physician of the prison, found him there, shivering and suffering from shock.

Another hero was "Big Jim" Morton, a celebrated Cleveland bank robber. He was at liberty in the yard when the fire broke out. He rushed into the cell block and began pulling out half-dead convicts. After he had carried out a dozen he said to one of the prisoners working over the men in the yard, "Guess maybe they ought to parole me for this. I've served the state well tonight."

Before he was through he rescued at least twenty men, and then his lungs filled with smoke and he staggered and fell. He was dragged out by another prisoner and revived.

There were others, too. Frank Ward, an ex-policeman, found a certain regeneration in the fiery holocaust. With an axe he hacked away at one cell door after another, releasing 136 men in all. Outside in the yard Ben Rudner, who was serving a life sentence for murder, saved life after life by operating a pulmotor. Nearby, convicts Ben Johnson and Aaron Friedman, both thought of as incorrigible criminals, handled the bodies brought from the prison, placing them tenderly in long, straight rows. A black prisoner in the yard found a rope and hook. After many attempts to throw the hook into a barred window of the upper tier, at last he succeeded and climbed the rope in an effort to reach the doomed prisoners. He failed and had to come down.

There was special irony in some of the deaths. Oren Hill, once a prison guard, was serving a three year to life sentence for helping John Whitfield, a notorious convict, to escape. He was himself held in his cell too long by men who had been fellow guards—and suffocated to death.

Another man who died, Garland Runyon, had been admitted into the prison only that day for abandoning children.

Some of the most undisciplined prisoners became the greatest heroes. Company E was a group of men considered to be so intractable that they were isolated so they could not breed discontent and rebellion. In their time they had fomented riots and even started fires in attempts to escape—but now they worked with guards to rescue prisoners from their cells.

Despite such heroic efforts, the flames spreading through the cell block continued to claim victims. Two prisoners, working with chisels on the welded and heat-warped locks, found themselves suddenly and irrevocably trapped. Rather than be devoured by the monstrous fire, they cut their own throats.

Another convict, seeking to escape outside the cell block wall by lowering himself on a rope, was the victim of a freak accident. The rope slipped, wound itself around his neck, and hanged him. His body dangled grotesquely from the cell window for hours as the fight to stem the fire went on.

One prisoner, who had escaped to the yard and then gone back inside to rescue others, suffered painful burns on his arms and hands as he dragged out men with flaming clothes and hair while beating out the fire with his hands. Another fought suffocating smoke for half an hour to pull a man out—and then discovered he had rescued his brother.

A fireman described conditions inside the burning cell block to a newspaper reporter later. "It was hell in there," he said. "While we were trying to cut through the steel the trapped prisoners climbed up the bars of the cells pleading with us to save them. We could hardly see through the smoke. We were driven back and these men died before our eyes. They were overcome toward the end and did not scream, so I think they were unconscious by the time the fire reached them."

In all this confusion only one inmate escaped to freedom—a man named Michael Dorn who donned the white uniform of a hospital intern and casually walked away. That only one escaped was no accident. From the first, officials in Warden Preston Thomas' office had been wrestling with a perplexing two-point problem. First, they had to rescue as many men as possible; second, they had to make sure there was no wholesale jail break.

To handle the first, Miss Amanda Thomas, daughter of the warden, gave the alarm that brought an army of firefighters and every piece of fire fighting equipment in Columbus to the stricken prison.

To handle the second emergency, one hundred fifty Columbus policemen were pulled off other duties and hurried to the penitentiary. The 116th Ohio National Guard Infantry and six hundred Federal troops from the garrison at Fort Hayes, along with 200 regular prison guards, moved in on the prison. Doctors and nurses from miles around converged on the scene of tragedy—as well as twenty thousand milling spectators.

But such is the perversity of human nature that not even the firemen, who arrived first, were permitted to do their work unhampered. In the spacious yard nearly four thousand prisoners had congregated. As they watched their mates being dragged from the burning building, they first grew sullen and then openly rebellious.

When Fire Chief Al Nice had the gates of the prison opened, the firemen entered, dragging hoses and carrying axes. To their surprise they were greeted by a jeering mob of prisoners who barred their path to the burning cell block.

The firemen halted in amazement, and Fire Chief Nice started to reason with the prisoners. But they mumbled angrily and edged toward the firemen. Nice, seeing the temper of the mob, discreetly withdrew his men. Outside the wall he gave vent to his disgust.

"I will not send my men in there unless they are given adequate protection," he told prison officials firmly.

Precious minutes slipped by and the fire grew alarmingly. At last the National Guardsmen and Federal troops, aided by Columbus policemen, moved in on the 4,000 mutinous convicts. They went in with bayonets gleaming and, as the inmates fell back, the firemen followed. At last the fight against the blaze began.

But the rebellious prisoners were not to be quelled so easily. Perhaps they saw in the fire and confusion a chance for complete escape, for many of them acted as if they wanted the entire prison to burn to the ground. For four long hours, as firemen fought the stubborn blaze, convicts in the yard sabotaged their efforts. Hoses, playing on the fire, would suddenly go dead—sliced in two by inmates armed with knives. More hoses would be brought into action, and these would be cut too. At length the hoses had to be guarded by troops, which meant that there were fewer to police the swarming hoard of antagonistic men.

In one instance firemen were washing down the blackened building with hoses when rocks and debris descended on them. Those holding the hose wavered, and the mob cheered wildly.

Another shower of rocks hit the firemen, knocking several to the ground. Troops started to move in, but this time they were not needed. The firemen, struggling to their feet, turned the high-powered hose on the rioters. The force of the water toppled the convicts like tenpins and stopped further harassment.

Firemen had hardly dealt with this situation when a group of prisoners attacked a fire truck. They had found a blazing blanket somewhere and, shouting with excitement, pushed the blanket under the truck. This time troops drove the rioters back with bayonets that glowed weirdly in the orange-red light of the flames.

On another occasion Fire Chief Nice, attempting to connect a hose to a hydrant, was set upon by several convicts. Before he could defend himself, they had picked him up bodily and hurled him against a wall.

The unruly mob even set other fires to divert the firemen from their main task. About 300 yards from the burning cell block was a woolen mill. Several hardened criminals left the main throng being held at bayonet point in the yard and raced toward the mill with pieces of burning wreckage. Before they could be stopped they had set it afire, and several firemen had to be diverted from fighting the cell block fire to put it out.

Another prisoner set fire to the Catholic chapel, using one of its candles as a torch. Firemen also extinguished this blaze before major damage was done. When a gate was opened to admit an ambulance, twenty convicts who tried to force their way out were driven back by guards with riot guns. Finally Warden Thomas was forced to issue a "shoot to kill" order, and this effectively quelled the disturbance.

By midnight all recalcitrant convicts were at last subdued. The huge penitentiary, by this time, looked as if it were under siege. The walls bristled with machine guns. Soldiers guarded each exit. Federal troops began to herd rebellious inmates into portions of the prison that had escaped the fire. At length, after five weary hours, the excitement died down—and the angry flames in the gutted cell block were finally extinguished.

Prison officials, accompanied by firemen, went into the charred building to survey the damage and were shocked by what they saw. The inside was a glowing hot mass of twisted steel and smoldering debris. On the upper tier of the ravaged building were 168 men, crushed and burned by the fall of the blazing roof. Cinderized corpses were everywhere. Unburned checkerboards, boxes of candy, opened books and playing cards were scattered about in wild profusion, showing how many had spent the last moments of their lives.

The six-tiered cell block had housed about eight hundred inmates. A total of three hundred twenty were dead, lying in long silent rows in the yard or crumpled in the ruins of the building. Some two hundred thirty were injured.

Not far away, at the Ohio State Fairgrounds, a combination morgue and hospital was being hastily established in a cattle barn. Trucks and ambulances carrying dead and injured moved into the area under heavy guard, like vehicles returning from a battle front. Physicians were on hand, and weeping women moved among the long line of corpses identifying bodies. All but a few, who were beyond recognition, were eventually claimed.

In Warden Thomas' office a grim scene took place. State Fire Marshall Ray Gill looked gravely at the distressed warden, who sat slumped at his desk.

"How do you think the fire might have started?" he asked.

The warden shook his head. "I don't know."

"Do you think," said Gill softly, "that the fire might have been deliberately set?"

Warden Thomas raised his head. He looked haggard and careworn.

"I don't think so."

"Some of your guards do."

"I don't," insisted Thomas.

Gill sighed. He took from a bag a pile of charred clothing, exhibiting it to the warden.

"I'm sorry to spoil your illusions about your men," he said.

"But these were found beneath a staircase in the scaffolding along side the west cell block. Obviously, someone piled them there—and set them afire. The scaffolding formed a perfect ladder for the flames to climb to the roof of the building."

A further, more detailed investigation proved Gill correct. The incendiary origin of the fire, however, did not take prison officials or the State of Ohio off the hook. A Board of Inquiry headed by Governor Myers Y. Cooper, after sifting all evidence, made it quite plain that the State and certain prison officials were basically responsible for the raging inferno—and the high death rate. They found that in refusing to give up the keys to the upper tier cells, Guard Thomas Watkinson had signed the death warrants of one hundred sixty-eight men. Watkinson insisted he was not to blame—that he had simply been carrying out the orders of his captain, John Hall. But Hall denied this and Watkinson was suspended by the warden.

Warden Thomas also came in for his share of the blame. It was found that he had placed a seventy-one-year-old deputy, J. C. Woodward, in charge of the institution after the fire broke out while he took a position outside the walls to see that no one escaped. This was considered a dereliction of duty. It was also found that prison personnel had never been instructed on what to do in case of a fire, nor had there been a fire drill within the memory of anyone working at the prison.

The inquiry also revealed that the first alarm received by the Fire Department had come from a box outside the prison walls—indicating much delay in reporting the blaze. Also, there was the surprising testimony that the main cage door leading to the cells, which usually was unlocked, was found to be locked during the holocaust—a strong indication that someone had locked the doors *after the fire started*.

Lastly, of course, the overcrowded condition of the prison—4,300 prisoners in accommodations for 1,500—contributed greatly to the high death toll.

During the dramatic inquiry, Fire Chief Nice reported,

"If the guards had not been so anxious to keep the men penned up, ample opportunity would have been present to save them all." And an outraged Catholic chaplain, the Reverend Albert O'Brien, said that "the disaster was a crime on the part of the State—a greater crime than any of those dead boys ever committed against the State."

Newspapers all over Ohio took up the cry. "Frantic men suffocated like vermin behind their steel bars," reported the *Ohio State Journal*. "Hundreds screamed, cursed, prayed, tore at their bars, begged to be saved—and died."

"Responsibility rests squarely on the State," said the *Columbus Evening Dispatch*. "For many years successive legislatures have dawdled over the prison problem while defenseless human lives remained in jeopardy."

And the *Cleveland Plain Dealer* stated: "The State must abandon a policy of neglect and indifference. The cries of men behind steel bars, held in a vise for creeping flames to destroy, are ringing in Ohio ears. The State is more cruel than we believe if the cries are unanswered."

Almost a year after the fire two convicts admitted having set the blaze as a protest against being forced to work on the scaffolding. They had poured oil over the scaffolding and over the pile of garments and then ignited this with a candle stolen from the chapel. Pleading guilty, they were given life sentences. Even so the State of Ohio shared the major part of the blame for the conditions that turned the makeshift fire into a major tragedy.

But the most ironic twist of the fire was the fact that, despite the great loss of life and the human suffering it caused, the cost of damage to the west cell block was only $11,000.

9

The Great Zeppelin Disaster (1937)

ONE of the clearest and sharpest memories of my boyhood was the moment that I stood in the streets of Detroit and watched a huge silvery dirigible hover majestically over the city. This was in the 1930s, and I do not remember if it was the *Graf Zeppelin* or the *Hindenburg* that I saw. What matters is that I saw it—and that no one born after the spring of 1937 has ever seen a mammoth airship of this kind in flight and probably never will.

For the day of the large passenger-carrying dirigible ended on May 6, 1937, in a fiery crash of the German zeppelin *Hindenburg* at Lakehurst, New Jersey.

During the first thirty-seven years of the twentieth century the dirigible was considered the coming thing in air transportation. The belief that the lighter-than-air dirigible was the answer to man's attempt to conquer the skies seemed to be based on firm footing. The cigar-shaped monsters could carry great weight for long distances. The ships were pleasant for passengers, they were durable, and they were able to travel in conditions of poor visibility that would ground heavier-than-air craft. Besides, airplanes before and for some time after World War I were incapable of flying more than a couple of passengers at a time, while dirigibles could accommodate large groups in comfort.

The era of the dirigible began on July 2, 1900. On that day the first rigid airship made its initial flight from a floating hangar on Lake Constance near Friedrichshafen, Germany—three years before the Wright Brothers made their first rickety flight at Kitty Hawk. The new aircraft was the first of the zeppelin-type airships, consisting of an elongated, trussed and covered frame supported by internal gas cells, and it was designed by a retired German army officer with the imposing name of Count Ferdinand Adolf August Heinrich Graf von Zeppelin.

Beneath the lighter-than-air craft were two external cars, or gondolas, one forward and a smaller one near the stern. Each contained a sixteen-horsepower engine geared to two propellers. A weight capable of sliding beneath the ship could be controlled to either raise or lower the nose, and there were rudders that provided horizontal control. The craft could attain a speed of close to twenty miles an hour. It was immediately hailed by the world as an important achievement in air transportation.

By 1910 Count von Zeppelin was running a regular schedule between Lake Constance and Dusseldorf, some 300 miles away. By 1913 he had completed sixteen hundred flights and had transported thirty-seven thousand passengers without mishap.

Then World War I burst upon the scene and Germany turned the rigid airships—by that time called zeppelins or zeps—to use as long-range bombers. On the night of May 31, 1915, the first zeppelin flew over London and dropped hundreds of incendiary bombs on a surprised and terrified city.

The attack was hailed as a great success by the German High Command, but it was not long before the kaiser discovered that the German airships were not going to have it all their way in warfare. A week later another raid was attempted and, as the airship tried to return to Germany, it was shot down in flames by a British pilot. This act convinced the Germans that the zeps were not suited to offensive warfare, since the hydrogen that buoyed them was highly inflammable.

By the time the war ended Count von Zeppelin had died and was replaced by Dr. Hugo Eckener, a zealot who strongly believed that the big airships, while of questionable value in war, were suitable for carrying passengers in peace time. Other air-minded individuals in other countries thought so too, and the United States, Italy, and Great Britain all began to build the big zeps.

At best, the dirigibles had a checkered career marked with impressive successes and frightful disasters. In the summer of 1919 the British dirigible *R 34* made a flight across the Atlantic and back—the first ever. Three years later a German zeppelin flew from Germany to the United States. In 1926 an Italian dirigible reached the North Pole.

In September, 1928, the Germans, who were the major producers of dirigibles, introduced the *Graf Zeppelin* to the world. This great ship was the first to inaugurate regularly scheduled passenger service across the Atlantic, and by the time it was decommissioned in 1937 it had made five hundred ninety flights, including 144 ocean crossings, and flown more than a million miles. It had also accomplished a spectacular round-the-world flight, covering 21,500 miles in twenty-one days.

But the dirigibles' record of achievement was marred by failures. The United States Navy dirigible *Shenandoah* crashed into the sea in 1925 with a loss of fourteen of her crew.

In 1930 the British airship *R 101* crashed in France, killing forty-seven people. Three years later another U. S. Navy ship went down in a storm and killed seventy-three. But these disasters occurred on military airships, and it was the boast of dirigible enthusiasts that there had never been an accident or a death in passenger service.

That boast came to an abrupt end in 1937.

The airship was the *Hindenburg,* first flown in 1936 and advertised as the biggest and safest dirigible ever built. It was the pride of Adolf Hitler's Nazi regime. Never before had the world seen such an airship, but the *Hindenburg* was destined to make only a brief splash on the pages of time and then die in agony, not only tarnishing the gaudy reputation of the Third Reich but bringing to a precipitous end the era of the great airships.

The *Hindenburg* (known by the code number of LZ-129, signifying the 129th zeppelin) was 146 feet high and 803 feet long—93 feet longer than the biggest battleship then in existence—and she was powered by four 1,100-horsepower Daimler-Benz engines. She had a maximum speed of 84 miles per hour, and her silver envelope carried 7,300,000 cubic feet of gas—in this case hydrogen, the colorless, odorless, and highly inflammable chemical element that was to prove her undoing. She had a cruising range sufficient to make a two-way crossing of the Atlantic ocean without refueling.

The passenger area of the *Hindenburg* was the ultimate in comfort, equaling in luxury most of the ocean-going passenger ships of the time. It was complete with a dining area, individual staterooms, bar, promenade deck and other attractive accouterments. On the portside A Deck was the spotless dining room which, between meals, served as a lounge. The lounge area was completely windowed so that passengers could sit in comfortable chairs and gaze up at the sky or down at the sea. Those wanting to know the exact position of the ship during a crossing could consult a chart with a small flag marking its location. A Deck also housed twenty-five staterooms down its center, each with two berths. On the starboard side of the deck were a music room, a library, and a writing room.

B Deck, directly beneath, was smaller because its observation windows were placed at a sharp angle permitting one to look directly down at the sea. It contained a bar and smoking room, as well as cabins for the ship's officers.

The *Hindenburg* seemed to be as much an airborne palace as it was possible to create.

The use of hydrogen to inflate the *Hindenburg* was a necessity. The Germans had no helium, a safer gas, and because Hitler was already rattling sabers and threatening war, the United States refused to sell him any. Well aware of the danger of fire, however, the Germans had taken elaborate precautions to eliminate the possibility. No passenger could smoke anywhere but in the smoking room, not even in his own stateroom. In fact, all matches and cigarette lighters were confiscated at boarding time, stored away, and returned at the end of the trip. Entering or leaving the smoking room was a complicated procedure that involved going through two doors. A passenger wishing to enter rang a bell that summoned a steward, who unlocked the first door. This door was closed behind the passenger before the second door was opened. In the room itself an electric lighter was available, but no matches. Even the air inside the room was maintained at a higher pressure than that outside to repel any stray hydrogen. When a passenger left the room he was required to close the inner door behind him and undergo scrutiny by the steward who, before opening the second door, made certain no lighted cigarettes were being carried outside.

Other safety measures demanded obedience from passengers and careful inspection by keen-eyed guards. Three catwalks in the ship were covered with rubber, and those walking along them had to wear sneakers or felt boots to prevent static or sparks. Crewmen whose duties required that they enter the topside area between the billowing gas cells wore asbestos suits with no buttons or metal.

These precautions, plus the Germans' penchant for demanding absolute compliance from both crew and passengers, gave the *Hindenburg* a reputation as the safest airship ever built. It was considered so safe, in fact, that the famous Lloyds of London insured it for 500,000 pounds at a low interest rate of five percent.

The *Hindenburg*—twice as long as the *Graf Zeppelin*—took to the air for the first time on March 4, 1936, for a trial run over Friedrichshafen. She maneuvered over the city for three hours and then came in for final adjustments and a last check before being placed in passenger service.

At 4:00 P.M. on May 6, 1936, the *Hindenburg* took off from Friedrichshafen on her first journey to the United States. She carried fifty-one passengers and a crew of fifty-six, and she made a smooth and trouble-free crossing of the Atlantic. When she floated over New York, proudly sporting Nazi swastikas on her fins, the city gave her a rousing ovation, and when she completed the trip at Lakehurst, New Jersey, her log showed that she had negotiated the crossing in sixty-one hours and thirty-five minutes. During the summer season of 1936 the *Hindenburg* made ten round trips between Germany and the United States carrying a total of 1,002 passengers. When winter came, she shifted her operations with the weather and made several trips to South America.

Big things were planned for the giant zeppelin in 1937. The *Hindenburg* was scheduled to make eighteen crossings, with only one-day stopovers at Lakehurst, thus providing a convenient shuttle service between the two countries that her owners felt would attract several thousand passengers.

It was Sunday, May 2, 1937, when the *Hindenburg* left Friedrichshafen for its first trans-Atlantic crossing of the new season. She was under the command of Captain Max Pruss, who had joined the German Zeppelin Transport Company in

1911 as a clerk and had become one of the best dirigible commanders in the business. Although he had flown the *Hindenburg* before on several flights to South America, this was the first time he had commanded the ship on a trip to the United States. With him on the flight was Captain Ernst August Lehmann, former commander of both the *Graf Zeppelin* and the *Hindenburg* and one of the most respected zeppelin commanders Germany had produced.

The trip proved to be slower than expected. Early in the flight, as she passed over the North Sea, the *Hindenburg* was beset by thunderstorms and headwinds. Still, the massive ship passed through the turbulent air with remarkable smoothness. At dawn on Thursday, May 6, she passed Nova Scotia, and before noon she had reached Boston. By three o'clock the silver ship was over New York, and the sight of the majestic dirigible virtually stopped traffic, as it always did. By the time she had headed for her mooring mast at Lakehurst, New Jersey, she was hours overdue.

Captain Pruss was disappointed that, on her maiden trip of 1937, the *Hindenburg* would arrive so late, for he was aware that her ambitious schedule for the year was already endangered. But when weather reports coming in from Commander Charles E. Rosendahl at the Lakehurst Air Station indicated that thunderstorms were due to hit the area within hours, Pruss decided to stop worrying about the schedule. He was a prudent man, determined not to take chances with the big zeppelin. He would cruise outside the turbulence for hours, if need be, rather than risk a mooring in a storm.

For three hours, from four until seven, the *Hindenburg* circled the storm's perimeter, once even flying over Atlantic City to amuse those aboard. Then, about 7:00 P.M., word came from Commander Rosendahl that conditions were favorable at Lakehurst. With the exception of a light rain, the storm had passed over the area, the wind had died to six knots, and visibility was excellent.

As the *Hindenburg* headed for a landing, activity on board increased. Stewards began to pile luggage near the exit; passengers crowded around the windows, peering down on the waiting throng below, trying to spot friends and relatives on hand to meet them. The passengers came from all walks of life: Joseph Spah, an acrobat using the stage name of Ben Dova, who had just completed an engagement at the Wintergarten Theater in Berlin and was on his way for an appearance at New York's Radio City Music Hall; Peter Belin, a Sorbonne student returning to his home in Washington, D.C.; John Pannes, a representative of the North German Lloyd Line in New York, and his wife; Hermann Doehner, owner of a wholesale drug firm in Mexico City, his wife, Matilda, and three children; Birger Brink, a newspaper correspondent for the Stockholm *Aftonbladet;* Mr. and Mrs. Leonhard Adelt, both authors, who planned to write a book entitled *Zeppelin;* Clifford L. Osbun, an export sales manager for a Chicago firm; James O'Laughlin, also a Chicago businessman; Philip Mangone, a clothing manufacturer out of New York; and others that included a German merchant, a Swedish editor, a San Francisco exporter, and the European representative of a New York advertising agency.

By this time those who had been waiting many hours for the arrival of the ship were weary and bored. At some distance from the landing circle, they gathered in a restless throng. Among them were the wife of the acrobat, with their three children; the Sorbonne student's parents; Robert Seelig, Murray Becker, and Larry Kennedy, all newspaper cameramen; Mangone's eighteen-year-old daughter, accompanied by his partner in the clothing business; George Willens and his son, Harry, who had booked passage on the return trip to attend the coronation of King George VI in London; Herbert Morrison, a radio announcer from station WLS, Chicago, on hand to broadcast a description of the *Hindenburg*'s arrival; Charlie Nehlsen, Morrison's engineer; many others who were there simply to get a look at the great zeppelin, and the 200 men in the ground crew whose job it was to help the *Hindenburg* to a safe mooring.

Although the landing was considered routine, airships were still rare enough to be news. There were six newsreel companies ready to film the mooring, correspondents from papers in New York and other nearby cities, and an Associated Press reporter poised to tap out a story of the landing that would be printed across the nation.

On the ground, Lakehurst Commander Rosendahl watched closely as Captain Pruss guided the mammoth ship to what looked like an easy landing. The ship approached from the south and, as she neared the mooring tower, made a sharp right turn that provoked an uneasy murmur from the watching crowd. Afterward, there was some discussion about Pruss's handling of the ship. Some thought that the abrupt turn was a clumsy maneuver, but Rosendahl was of the opinion that the tight turn was necessary to counteract a sudden change in the wind. In any case, the *Hindenburg* moved in slowly, and, at exactly 7:21 P.M., dropped the first landing rope from a height of about one hundred fifty feet. A minute later a second rope hit the ground. Both were taken by members of Frederick Tobin's ground crew and fastened to two small railway cars that operated on a circular track around the mooring tower.

In the control car of the ship Pruss and Lehmann smiled at each other. It was a routine landing after a long journey, the culmination of another safe trip that would be added to zeppelin history. In fact, radio officer Willy Speck was already tapping out the news that the *Hindenburg* had just landed safely at Lakehurst, and Herb Rau of the Standard News Association was phoning his night editor to report the same thing. And at the edge of the field Herbert Morrison was talking into his hand-held microphone, describing the scene.

"Passengers are looking out of the windows and waving," he said. "The ship is standing still now."

She was, indeed, standing still. She hovered with lofty

pride seventy-five feet above the ground, her silver nose nudging the mooring mast, engines reversed by order of Pruss and turning over silently, holding the ship in position. Radio announcer Morrison spun off some superlatives to describe the magnificent sight.

Then, at 7:25, everything changed. It was sudden, shocking, and totally devastating. At first there was only a tiny, flickering flame—some people said two of them—and seconds later the huge silver envelope of the ship was ablaze.

Almost simultaneously with the appearance of the tiny flames there were two explosions. To some they sounded like firecrackers or shots from a rifle. Others described them as thumps or dull thuds; still others insisted they were terrific blasts that shook the dirigible.

Directly beneath the belly of the ship at the moment of disaster was W. W. Groves, an engineer whose job was to check the airship's copper tubing. As he looked up he saw a small spark near the tail end of the ship. In seconds fire crept up the gas-filled envelope and the whole back of the ship was swept by lurid red flames.

At the same moment, Rosendahl saw a tongue of flame at the top of the hull. He was experienced enough in the way of dirigibles to know that this meant imminent disaster.

"My God!" he cried. "It's on fire!"

As the flames quickly consumed the entire stern of the giant dirigible its tail began to drop toward the ground. Someone in the ground crew screamed, "Run for your lives!" and the 200 men who had expected to guide the ship into safe mooring scattered. Inside the passenger gondola, people were knocked off their feet by the sharp pitch of the floor beneath them as the stern sank toward the ground. They knew, from the sudden brightness of the sky around them and from the warmth of the flaming envelope, that the ship was afire. Panic-stricken men and women began jumping from the ship. Some were killed leaping from too great a height. Others waited until the tail end of the ship neared the ground before jumping. Most of these were able to flee to safety, but some were immediately turned into torches by the flaming debris that fell over them in a fiery blanket.

On the field, radio announcer Morrison was "talking" the big ship into its mooring when the fire started. For a split second he stopped speaking, staring in awe at the flames. Then, half hysterically, words tumbled from his mouth, breathless, disconnected.

"It's burst into flames . . . it is burning, bursting into flames . . . it's falling on the mooring mast . . . it's crashing, crashing . . . this is terrible, terrible . . . this is one of the worst catastrophes in the world!"

In the control car forward, Captain Pruss became aware of imminent danger when he felt the ship jar slightly from one of the explosions. He was not in a position to see the burning tail and for a moment he was puzzled. Then he heard a cry from radio officer Speck.

"The ship's burning!"

As the tail of the ship began to dip, the captain's first thought was to drop the rear ballast to keep the ship on an even keel. But then he realized that such action would doom those passengers and crewmen in the stern, and he permitted the tail to drop rapidly so that they would at least have a chance to jump to safety. As the tail sank, the bow shot 500 feet into the air, and passengers and crewmen were tumbled on slanted decks. Still, many scrambled to their feet and jumped from the windows to the wet, sandy ground.

Joseph Spah, the acrobat, was standing at a portside window on the forward promenade deck trying to locate his wife and children in the crowd below when the fire broke out. Knowing that there was no chance for him if he stayed in the ship, Spah struck the window with a movie camera and, luckily, the entire pane fell out.

The bow of the ship began to rise in the air as the stern

dropped and Spah estimated he was more than a hundred feet from the ground. He knew that a jump from that height would be disastrous, and he was aware that the entire ship would sink to the ground eventually.

Climbing out of the window, Spah hung on to the ledge with one arm, ready to jump when the gondola dropped closer to the ground. He was thankful for his strong arms, developed from his profession, and that one of his favorite acrobatic stunts was to hang from a swaying lamp post with one arm.

Two men climbed outside the window with Spah, but they could not hold on. One man slipped away and fell more than one hundred feet, hitting the ground with a thud, bouncing, then sprawling inert. The second man held on a little longer, but then fell with a scream.

The bow of the ship rose higher and Spah wondered if he could hold on for the eternity it would take for it to descend again. Finally, as his arm tired, the nose of the ship began to fall. When he was about forty feet from the ground he let go. He hit the ground forcibly, fell to his knees, and crawled painfully away from the searing heat that was dropping down on him from above.

In the crowd at the edge of the field, his wife fought back panic. Gathering her children around her, she ran to one of the nearby airplane hangars, thinking the children would be safe there while she searched for her husband. When they arrived a soldier entered the building, shouting, "Spah!"

She identified herself, and the soldier told her that her husband was alive.

She found him sitting on a bench in the hangar, head bowed, blackened by the soot and smoke, but otherwise in good condition. For a long time Spah could not recall just how he had escaped from the burning ship.

Commander Rosendahl stayed at his post until the last. The stern of the ship was on the ground, the forward section poked into the sky at a forty-five-degree angle. The flames reached the center of the airship, then spread quickly to the bow. Rosendahl ran from under the falling wreckage as the bow began to descend.

As the great airship crashed to the ground, people poured from the gondolas like ants forsaking an anthill. They stumbled through flame, smoke and the burning skeleton of the ship in an effort to get away from the hell of the holocaust. Some escaped. Those who perished were a pitiful sight. One man was seen fleeing the wreckage with a tongue of flame pursuing him. He fell, got up and staggered a few more feet, then fell again. He crawled a short distance and finally lay still. Later, rescuers found that his body was charred black. Another man, with clothing and hair burned away, ran from the wreck chattering excitedly in German. A rescue worker took his arm and guided him to an ambulance, but he died as he was being put into it. Rescuers also found a girl so badly burned that when they tried to lift her body the flesh came off in their hands.

Author Leonhard Adelt was standing with his wife at an open observation window. He heard a dull thump from the stern of the ship and glanced around to see what was happening. He could not see flame, but there was a red glow. Instinctively he knew the ship was afire.

Glancing down, he estimated that the ship was more than a hundred feet in the air. The thought crossed his mind that he could return to his cabin, get bed linen and make a rope of it on which to descend. But there was no time for that. Suddenly the entire ship crashed to the ground. The impact threw both Adelt and his wife off their feet.

Scrambling upright, Adelt pulled his wife up with him. "Through the window!" he shouted. The two crawled from the window, their feet touching the sandy ground that seemed so firm and safe but was only the beginning of their agony. Black clouds of smoke blinded and choked them as they stumbled through still-blazing wreckage. Several times Adelt pried red-hot wires apart with his hands, feeling nothing.

Suddenly Adelt became aware that his wife was no longer with him. Glancing behind him through the smoke, he saw her lying on the ground. He rushed back, pulled her to her feet, and shoved her ahead of him. To his relief he saw that she was running again. Adelt fell once as he ran after her, but struggled to his feet and went on. When he finally reached safety, he turned around and saw the flaming debris and blackened skeleton that had been the *Hindenburg*. Without knowing what he was doing, he started to return to what was left of the ship. He wondered later if it had been an urge for self-destruction or a feeling that he must aid others, but in any case his wife took his arm and led him away to safety.

Clothing manufacturer Philip Mangone's first intimation of danger came when the airship shook violently enough to throw him off his feet. Getting up, he tried to open one of the observation windows. When he found it locked he grabbed a chair and broke the glass, then crawled through the window and hung there for a moment. But already the ledge was growing hot under his hands and he had to let go. He fell about thirty feet to the sand.

Looking up, Mangone saw the burning ship directly above him. Fragments of the flaming fabric and red hot wires were descending on him, and then, with a crash, the entire skeleton of the dirgible fell over him, pinning him in molten wreckage.

Mangone knew he might well die in the sheer heat of the debris. Frantically he started to burrow into the soft sand. Within minutes he had dug a trench beneath a mass of wreckage and had crawled through it to safety. Exhausted, he limped away with his coat on fire and his hair singed from his head.

Rescue workers immediately grabbed Mangone, beat the

flames from his coat, and tried to lead him to an ambulance. Mangone shook them off.

"I won't go to a hospital," he said. "My daughters are here. I have to find them."

A moment later his daughter Katherine found him. His face was red from the flames and his hands were burned. She persuaded him to take the ambulance to the hospital.

Peter Belin, the Sorbonne student, was near Mangone when the explosion shook the ship. He, too, was thrown to the deck. Like Mangone, he broke a window with a chair, leaped out, and when he hit the earth ran to safety. His parents, who had been convinced that he was dead, received him joyfully.

Matilda Doehner, wife of the wholesale druggist from Mexico, was at a starboard window when the dirigible dipped. With her were her three children—the only child passengers aboard the *Hindenburg*. Her husband was nowhere to be seen and Mrs. Doehner had to act. With the ship dropping to the ground, she decided to throw her three children from the window. The oldest, Irene, pulled away from her, however, and disappeared in a search for her father. Mrs. Doehner calmly threw the other two children from the burning ship and then jumped herself. All three were led away from the fiery wreckage by two rescuers.

Probably the most fantastic escape of all was that of Werner Franz, a 14-year-old cabin boy—who nearly died before a lucky break saved his life. Dizzy and reeling from the heat of the burning ship, Franz leaped to the ground. When he struck the sand he collapsed, almost losing consciousness. He glanced up, saw the flaming dirigible falling directly toward him and thought, *This is the end.* But fortunately a water tank overhead burst from the heat and spilled its contents over the boy like a protective cloak. The cold water revived him and he escaped with only minor burns.

In the control car Captain Pruss, Captain Lehmann, and ten crew members stayed with the ship until the last possible moment. The bow, which had risen into the air when Pruss purposely let the tail end of the ship settle, finally began to fall. At this point the fire had swept through the entire ship, and the descent of the control car to the ground was rapid. No one left the car until it hit the sand, but as they did so molten wreckage rained down on them.

Pruss staggered through the searing flames until he was beyond their reach, then beat out his burning clothes with his hands. When he was certain that he was out of danger, he rushed back into the smoldering ruins to rescue as many passengers as he could. Twice he dragged half-dead victims to a place of safety. But when he tried to return a third time, rescuers forced him into an ambulance.

Meanwhile, Captain Lehmann was in mortal danger. Having scrambled from the control car with his clothes on fire, he was discovered wandering around in a state of shock. As the rescuers beat out his flaming garments, Lehmann kept murmuring, *"Ich kann es nicht verstehen."* ("I can't understand it.") Horribly burned, Lehmann died the following day at the hospital. Pruss, however, survived.

One of the most graphic eyewitness accounts of the tragedy was written later by George Willens, who had expected to make the return trip to Europe:

> The weather was bad so the crowd at the field was small—probably less than two hundred people—many of them, like myself, having booked for the return voyage. I was busy with my movie camera—taking pictures here and there—when a spectator said, "There she is!"
>
> I looked and, sure enough, through the rain and mist the *Hindenburg* was approaching slowly, majestically from the west. It loomed suddenly out of the sky like a ship at sea in a fog. It was almost upon us before we knew it. The ground crew caught the lines she threw out and were plodding through the rain-soaked field with the big ship in tow. The crowd moved toward the mooring mast, some waving to friends on the ship.

Then, as I worked my camera, I saw a streak of flame shoot from the upper seam of the ship, back near the tail. "Strange," I thought, never dreaming that tragedy and disaster were seconds away. I kept working my camera. The flames, more intense in volume, continued to roll along the ship's seam. The man beside me clutched at my arm. "What is it?" he shouted in my ear.

Suddenly there was an explosion. Then the ship burst into flames and settled, like a stricken mastodon. From the crowd came, at first, a long-drawn sigh—then silence, except for the crackling of the flames and the scream of a woman here and there. I saw two men catapult out of the ship—several jumped—and the crowd scurried back as the heat of the flames, now enveloping the entire ship, became more and more unendurable. It was all so sudden—so unexpected—that no one knew what to do or what to say. It seemed as if no living thing could pass unscathed through that hell of flames. But, thank God, while many were fatally burned and many suffered grievous injuries, through the mercy of a Divine providence, most were saved.

Other eyewitness accounts were similar. Seelig, Becker and Kennedy, the newspaper cameramen, related what they saw.

Seelig said, "There was a noise that sounded like bullets coming out of the gondolas. I saw nobody jump, but I heard everybody on the ground screaming. The heat made my face tighten up."

Becker recalled:

I had my camera up to eye level when the ship burst toward the tail-end. The tail went first and the nose seemed to hang in the air. I saw no one jumping because I was so far back. I ran toward the ship and saw it enveloped in flames. In a fraction of a second there was nothing left but the skeleton. There wasn't much smoke. I saw a man walking toward me, assisted by two other men. He had no clothes on. I saw a woman lying on a stretcher. There were screams from men and women all over the field.

"It burst right over our heads," Kennedy related. "It flew

apart as if made of paper. Pieces of the fabric fell on us. I saw one fellow jump out, or maybe he fell out. I think he was a passenger. He lay moaning on the ground."

The breathless, hysterical accounts of other observers differed widely. A member of the ground crew, who was beneath the tail of the ship when she caught fire, said:

> She dropped her lines at 7:20 daylight time. There was a burst of flame and a loud report which appeared to be just aft of the port gondola. There was a loud shout, "Run for your lives!" The second explosion came about thirty seconds later.
>
> With several of the others I ran as far out of the circle as I could. I saw the ship just sink down and the flames go through it. The fabric burned away in just a few seconds. I turned back then with others to get as close to the ship as possible to pick up the survivors.

Alfred Snook, a dairyman making a delivery, saw the accident from a distance:

> I saw a spurt of flames from the dirigible. It seemed to come from the rear of the ship. Then there was a terrific explosion and the entire airship suddenly became enveloped in flames. The nose of the airship was jerked upward and then the whole flaming hulk plummeted to the ground where the wreckage was instantly enveloped in dense black smoke. I saw a lot of people rushing from the hangar, which was about one thousand feet from where the airship crashed. I did not see any life aboard the airship from the time it exploded until it struck the ground. I saw nobody jump from the ship and I saw nothing fall. A light rain was falling at the time, but I saw no flash of lightning and I heard no thunder.

Walter Cullen, a ticket agent for the American Air Lines, and Herbert Holson, his helper, were waiting on a truck to take the mail off the *Hindenburg*. Cullen said, "That blast was so terrific it blew both of us right off the truck. The fire seemed to go through the middle of the ship in an instant. Holson and I went to the ship as fast as we could run and the gondola was already on the ground when we got there. We climbed in and helped to pull several passengers out. But most of them took care of themselves."

Amazingly, the horrible conflagration was perhaps the briefest fire in history. Stopwatches on newsreel cameras later showed that only thirty-four seconds after the first spurt of flame at the rear of the ship, the burned carcass was on the ground, her 7,300,000 cubic feet of hydrogen consumed. In that short interval, thirty-six perished, including sixteen passengers and twenty crewmen.

The reason for the *Hindenburg* fire is still uncertain. Immediately after the disaster, speculations poured from every dirigible man in the world. Dr. Hugo Eckener, builder of the great airship, said the fire was probably caused by a hydrogen leak ignited by a backfire or static electricity. Commander Rosendahl and many others thought sabotage was involved. Other suggestions were that an exhaust spark might have ignited escaping gas; that the tight turn made by Captain Pruss as he approached the mooring mast broke a bracing wire in the hull and that the wire then cut open a gas cell, and that the landing line hitting the ground caused a spark that ignited escaping gas.

Both the United States and German governments made thorough investigations of the tragedy. The findings were combined in August 1938 in an official report by R. W. Knight, acting chief, Air Transportation Section of the Bureau of Air Commerce, Safety and Planning Division, in the United States Department of Commerce. Both governments considered a large variety of possible causes: sabotage, the presence of a combustible mixture of hydrogen and air through diffusion, the sticking of a valve, entry of a propeller fragment, leaking of a gas cell, major structural failure, ball lightning, brush discharge (also called St. Elmo's Fire), and many others. In most areas the two investigations agreed.

For example, both governments agreed that sabotage was

unlikely, ruling out incendiary bullets because this type of attack would have been easily detected and would not have caused an explosion, discounting the possibility of time-fuses or other releases from within the ship because the ship had been under close supervision at all times, and eliminating other possibilities, such as an attack from other aircraft, which would have also been detectable. The presence of a combustible mixture of hydrogen and air was discounted as a possible cause because the ventilation system could not have allowed a rich enough mixture to have formed through normal diffusion.

The sticking of a valve was a stickier question. The German report states that "it is not completely excluded" that one of the maneuvering valves may have become stuck during valving operations some minutes prior to the dropping of the landing ropes, and the American report lists the failure of the valves as a possibility, although it notes that there "is no testimony on the subject."

Both investigations agreed that there was no propeller breakage until the engines struck the ground. Another area in which both governments concurred was the leaking of a gas cell; both admitted that breakage of a steel brace wire could have punctured one of the rear gas cells. And both reports discounted the likelihood of major structural failure.

The conclusions reached in the reports were virtually identical. The German investigative commission stated that "in spite of thorough questioning of all witnesses, in spite of thorough-going inspection and search of the wreckage, evaluation of all pictorial documents giving testimony of the sequence of the fire, no completely certain proof can be found for any of the possibilities considered." It found that "the following explanation of the accident appears to be the most probable: During the landing approach a leak developed in gas cell four or five at the stern, perhaps through the tearing of a wire. Leaking hydrogen then entered into the space between the cell and the outer cover, causing an inflammable mixture of hydrogen and air to gather in the upper part of the airship." It was thought that ignition of the gas could have come from two sources—brush discharges due to atmospheric electrical disturbances, or a spark caused by equalization of tension between wet spots on the exterior and the ship's framework, the latter being better grounded than the covering after dropping the manila landing ropes. The "equalization of tension" theory was considered the most likely by the Germans.

The American report agreed that a leak occurred in the vicinity of cells four and five, causing a combustible mixture of gas and air. But it differed in not mentioning the "equalization of tension" theory, stating instead that "the most likely cause of ignition was brush discharge."

It is interesting to note that a more recent investigation of the *Hindenburg* disaster by A. A. Hoehling in his book *Who Destroyed the* Hindenburg? paints a highly convincing case for sabotage by a crewman disgruntled with the Nazi regime, concluding that the placing of an incendiary device near cell four is "more than probable."

Perhaps no one will ever know the truth about the spectacular demise of the *Hindenburg*. However, one thing is known for certain: the tragic end of the *Hindenburg* put the finishing touches to the dirigible as a means of transportation. She was the last great airship ever to carry a passenger.

10

Boston's Nightclub Tragedy (1942)

It was early winter, 1942. All across Europe, in northern Africa, and in remote areas of the Pacific, World War II was raging. The Russians had begun a counteroffensive against the Germans at Stalingrad and were pushing the proud Nazi armies into headlong retreat. United States and British troops were attacking the Bizerte-Tunis line in northern Africa, and just a week earlier the Japanese had been routed by United States forces from an island called Guadalcanal.

With the war turning subtly in favor of the Allies—but with a long way yet to go—Americans were being subjected only to the relatively mild discomforts of gasoline and coffee rationing. Still, the grim news of battles and monstrous casualties was depressing, and to take their minds off the war for even an hour or two, people often sought entertainment. Thus it was that the Cocoanut Grove nightclub in Boston was packed with the biggest crowd in its history on the tragic evening of November 28, 1942.

Owned by Barnet Wilansky, a prominent figure in Boston's nightlife, the Cocoanut Grove was the oldest and most prestigious nightclub in the city. Its fifteen-year history had been a checkered one. Its first owner was a prohibition gangster with more enemies than friends. After he was summarily dispatched by rivals, the club changed hands several times. But despite ups and downs in its ownership, in the early days it had featured top-ranking stars such as Texas Guinan, Helen Morgan, and Joe Frisco, and during World War II it was still considered the "in" place to go for an evening of solid entertainment.

Now, in the fall of 1942, it was enjoying an unprecedented boom. Service men home on furlough and war workers flush with overtime paychecks jammed it night after night. To accommodate extra-large crowds, the club had expanded its facilities.

The Cocoanut Grove occupied a one-story building between Piedmont Street and Shawmut Avenue in Boston's midtown theater district. It was a hodgepodge of rooms. The revolving door entrance on Piedmont Street led to a large foyer, which, in turn, connected with the main dining room, a spacious area of white-clothed tables, a dance floor, a bar, and a rolling stage used for floor shows. There was a double-door exit from the dining room to Shawmut Avenue, as well as a hallway exit that led to a service door. A new cocktail lounge had been added to the dining area only two weeks before tragedy struck. Except for the service door on Shawmut Avenue, the lounge itself had only one other exit, this one facing Broadway, a side street.

On the Piedmont Street side were two bars. One, part of the main dining area, was called the Caricature Bar; the other, down a few steps from street level, was known as the Melody Lounge. It was considered the "intimate" lounge where young men and their dates sat in semi-darkness at the bar or in the booths along the wall. The stairway from the main level was a narrow one.

On the fatal night of November 28 a diverse crowd of more than eight hundred jammed the Cocoanut Grove's various rooms. Among them were soldiers, sailors, marines and coastguardsmen in town for a weekend of merriment, along with the usual quota of out-of-town businessmen and Boston war workers. In addition, there were many football fans who had attended the Boston College-Holy Cross game at Fenway Park that afternoon. Holy Cross had won a major upset, and half of the fans at the nightclub were celebrating the unexpected victory while the other half were trying to drown their sorrows.

This, then, was the scene when fire started in the below-ground-level Melody Lounge . . .

Eight days before the fire, Lieutenant Frank J. Linney, a gray-thatched veteran of the Boston Fire Prevention Bureau, strolled into the Cocoanut Grove with an order from the Fire Department to inspect the premises. In the absence of owner Barnet Wilansky, he was conducted around the building by Wilansky's brother, James. During the inspection, only one thing seemed to arouse the misgivings of Lieutenant Linney. The club was decorated with artificial palm trees to give it the appearance of a grove of cocoanut palms, and Linney suspected that the flimsy-looking decorations might be highly inflammable. To test them, he deliberately held a match to five of the palm fronds. None ignited.

When the inspection had been completed, Wilansky asked, "Well, do we pass?"

Lieutenant Linney nodded. "I'd say conditions are good here," he replied.

Linney returned to his office and prepared a report. It read:

Subject: Inspection, Cocoanut Grove, 17 Piedmont Street.
Owner and manager: Barnet Wilansky.
Building: One story, occupies approximately 7,500 square feet.
I submit the following report of inspection made this day, November 20, 1942, and in my opinion condition of the premises is good.
A new addition has been added on the Broadway side, used as a cocktail lounge room, seating 100 people. No inflammable decorations.
Main dining room seating 400 people. Old tenement house next to new lounge room. Second floor of tenement house used as a dressing room. Third floor used as help locker room.
This building connects into hallways between lounge and main dining room.
Kitchen in basement, free from grease, hood over stove, underside clean. Cocktail bar in basement.
Sufficient number of exits.
Sufficient number of extinguishers.
Heat, using fuel oil, two 375-gallon tanks and coal.
Condition—good.

Lieutenant Linney had made a routine investigation and had detected nothing wrong. He left his office that day, pleased with a job well done. . . .

Three weeks before the fire Stanley Tomaszewski, a sixteen-year-old high school honor student and a football player, was hired as a bus boy. After two weeks he was promoted to bar boy, a move that pleased the ambitious young man. In his new job he worked from 4 P.M. to 1:15 A.M., setting up glasses at the bars and generally making himself useful to the bartenders.

On the night of November 28 Tomaszewski was working in the basement Melody Lounge under the direction of bartender John Bradley. He noticed that, although the Cocoanut Grove was always a busy place, it was particularly crowded on this occasion and he was hard-pressed to keep up with his various duties. Still, he liked his work. There was glamor in his surroundings, and occasionally he mingled with celebrities.

The most famous personage there on November 28 was Charles (Buck) Jones, the cowboy motion picture star. A longtime favorite of movie-goers, Jones had viewed the Boston College-Holy Cross football game from Mayor Maurice Tobin's private box and had come to the Cocoanut Grove that evening as the club's guest of honor. With him was an entourage of some 24 show business people: Martin Sheridan, Jones' Boston representative, and his wife; Scott Dunlap, one of Hollywood's leading producers of Western movies; Edward Ansin, president of the Interstate Theatres Corporation; Philip Seletsky, chief film booker for a string of Boston theaters; Eugene Goss, one-time associate of the famous Cecil B. DeMille; Moses Grassgreen of Universal Pictures; Bernard Levin of Columbia Pictures, and others. The distinguished group had choice tables on a terrace in the main dining room, from which they had an unobstructed view of the rolling platform where Mickey Alpert and his band played and the floor show was staged. The performers that night were club singer Billy Payne, the dance team of Pierce and Roland, acrobatic dancer Miriam Johnson, violinist Helen Fay, and a mixed chorus of young men and women dancers. . . .

It was shortly after 10 P.M. In the over-crowded Melody Lounge, where more than one hundred and thirty people sat drinking as they listened to ragtime tunes from the piano, a soldier and his sweetheart huddled in a shadowy corner booth. Overhead was a lightbulb that brightened the booth more than the soldier desired. Impatiently, he stood up on the seat and loosened the bulb until the light went out.

John Bradley, one of five bartenders trying to keep up with the demand for drinks, beckoned to Stanley Tomaszewski, the bar boy, and told him to put the light back on.

Tomaszewski grabbed a bar stool, approached the booth,

and told the soldier why he was there. But when he climbed up on the stool, he was unable to locate the bulb in the shadows. Carefully he lit a match and tightened the bulb.

As he got down from the bar stool, Tomaszewski shook out the match, dropped it to the floor, and stepped on it. What he did not realize was that, while he was tightening the bulb, the match had ignited one of the artificial palms near the ceiling.

What happened then was a flash fire in every sense of the word. By the time Tomaszewski descended from the stool, the flame had already leaped from the palm tree to some nearby curtains. Bradley, at the bar, came rushing over. Frantically, they tried to pull down the burning curtains and beat out the flames, but in seconds the entire corner of the Melody Lounge was ablaze. In terror, Tomaszewski fled to the basement kitchen and escaped from the building.

Customers in the overcrowded lounge, however, reacted more slowly. The fire was already out of control in the corner of the room when somebody shouted "Fire!" and people began to get to their feet. By this time a thick cloud of black smoke hung over the room.

Suddenly panic-stricken, the lounge's one hundred thirty occupants bolted for the single exit—the narrow flight of stairs leading to the street-level foyer. Many fell on the stairs and were trampled. Others, gasping from the thick smoke, fell to the floor. Those still on their feet jammed the stairway, and the narrow passage became a scene of horror as people feverishly tried to climb over each other.

They had little chance of escape. The fire leaped across the room and ate away at the decorations, the curtains, the bar itself. Screaming men and women, caught in the maelstrom of bodies at the staircase, were turned into human torches. The flames, unimpeded, leaped up the stairway. Only a handful of people made it to the foyer, pursued by flames.

Those in the main dining room were not yet aware of the

fire below. The floor show was about to start, and band leader Mickey Alpert was raising his baton to begin the national anthem, a standard patriotic practice in wartime. Singer Billy Payne was at the microphone. But the anthem was never played. At that moment a terrified young woman who had apparently escaped from the blazing Melody Lounge raced across the dance floor with her hair afire. And the next moment the flames from the stairs swept over the more than 400 incredulous patrons in the dining room.

Everything happened in seconds. The red horror of the flames spread rapidly across the ceiling of the dining room, raining sparks. The lights had been lowered for the floor show, and the fire produced a heavy, choking smoke. Inexorably the blaze moved on to engulf the new cocktail lounge.

People had little time to react to the terrifying onslaught of flames. They simply rose from their tables and rushed blindly for the only exit they knew—the revolving doors leading from the foyer to the street. Calmer heads tried to quiet the stampeding crowd, but there was no chance to do so. Women fell and were trampled. Stark fear had turned the patrons of the club into a mob with no other thought than to escape from a fiery death.

The result was catastrophe. At the revolving doors, those who had managed to make their way from the Melody Lounge over piles of corpses on the stairs converged with those from the dining room. The frightened multitude tried to push the doors both ways at once. They jammed, and a mass of squirming bodies piled up at the exit.

Next to the revolving doors was an auxiliary door equipped with a panic lock that was supposed to open readily to pressure from inside. But when the surge of people tried to push through this door they found that it had been bolted. In the smoke-filled foyer the first wave of people to reach the auxiliary door were crushed against it by the fighting crowd behind and could not locate the bolt.

To add to the horror, the lights of the nightclub suddenly went out. All the rooms—the Melody Lounge, the dining room with its Caricature bar, the new cocktail lounge, and the foyer—were plunged into darkness. The only illumination was from the flames dancing across the ceiling and the eerie light of human beings turned into pillars of fire.

While the dining room crowd was halted at the foyer doors, people in the new cocktail lounge had similar problems. The lounge was equipped with two doors, the service door and the customer exit on Broadway. Most of the patrons knew of only the customer exit, and as the fire spread quickly from the dining room into the new lounge there was a mad scramble for this exit. But the first to reach the door found, to their horror, that it *opened inward*—and before they could pull the door toward them the pressure of the surging crowd behind jammed it. More than one hundred bodies were later found piled in a tangled heap against the unyielding door.

About fifty patrons made what seemed to be an intelligent decision. With exit doors blocked and fire raging through the club, they tried to escape through the basement kitchen. But they, too, found themselves trapped. In the pitch blackness, and with heavy smoke seeping into the kitchen area, they were unable to find the exit. They milled about until most of them perished from smoke inhalation.

Only one of them escaped by the kitchen route. He was Charles W. Disbrow, Jr., a Boston insurance executive, who had been dining with his wife and eight friends in the main dining room. The group had just finished dinner when Disbrow noticed a disturbance at the dining room bar. Later he told the following story:

> I saw a man jump on the bar, which was raised above the main floor. Two belches of flame seemed to be chasing him. I grabbed my wife and said, "This is no place for us." I knew there would be panic and people had already started running away from the space at the bar. We made for the servant door and went down a flight of steps into the basement.

> There was no smoke in the cellar as we entered. There seemed to be a line of people down there, about two abreast, and they were going toward a dark corner. We went after them and then someone yelled, "We can't get out this way!" We began to fight our way upstairs again when smoke began enveloping the basement. I guess that smoke really saved us. Instead of trying to get back upstairs, we decided that we would have to grope around the cellar looking for a way out. Then the lights started to go out, and we kept trying to find an exit. We made our way into a storeroom but had to retreat again. Then we felt a current of fresh air through the smoke and followed it. The air was coming from an open window. I smashed the window and we climbed through.

Not all of the people who died on the main floor of the Cocoanut Grove that night succumbed to flames. Some never rose to their feet, dying of smoke inhalation before they could move. It was later suspected that the black smoke swirling through the club contained toxic gases from burning paint or decorative material.

Buck Jones, seated with his retinue on the dining room terrace as flames swept across the ceiling, rose and started for a door, but smoke overwhelmed him. He staggered and fell near the corner of the terrace. After the fire, rescuers noticed a man crumpled on the terrace wearing handsomely tooled leather cowboy boots. Of course it was Jones, alive but unconscious. He was rushed to Massachusetts General Hospital, where he died—the only celebrity casualty of the evening.

A woman named Joyce Spector was worried about a new fur coat she had checked at the cloak room. She made her way toward the counter but never reached it. Fire, surging up from the Melody Lounge, soared over her head, and the next moment she was knocked from her feet by the stampeding crowd. By some miracle she crawled on hands and knees toward a doorway and managed to squeeze through to safety before battling mobs blocked it.

John C. Gill, head of the Catholic Youth of Boston College Alumni, and his wife, Margaret, were having dinner in the main dining room when Gill noticed what he later described as a "vague flurry" in the direction of the foyer. With it came the sound of a falling table and the tinkle of glasses breaking. Gill turned in his chair, saw a great mass of flame belly up from the Melody Lounge and turned quickly to his wife.

"It's a fire. But be calm. Be calm."

Even as he uttered the words panic was all around him. Women screamed. Men cried out hoarsely. The unreasoning throng surged toward the Gill's table and knocked both of them to the floor.

Trying to protect his wife from the crowd hurtling toward the foyer exit, Gill covered her body with his own. Each time he tried to regain his feet he was knocked down, trampled, and kicked. Flaming debris began to descend on them, burning their faces and hands, setting their clothing afire.

The relentless crowd kept trampling them and Gill kept thinking, *They're going over us. They're going over us to freedom.*

All the while, Gill and his wife tried to beat out their scorched clothing. His hair was singed, skin was peeling from his face as flames tortured him, and his hands were burned.

Finally Gill managed to get to his feet. He pulled his wife upright and, with one arm, warded off the surging crowd. He saw men and women, already afire, coming toward him like torches. He thought he heard plate glass crashing and instinctively he guided his wife toward the sound, on the chance that somewhere in the mad holocaust somebody had broken a window. What happened then is unclear to him.

"One minute," he said later, "I was moving toward that sound of breaking glass, dragging Margaret with me. The next few seconds, or minutes, I couldn't say which, are blanks in my memory. I don't remember that we were borne through that little door (most likely the Shawmut Avenue service door from the cocktail lounge), but we must have been. I don't know whether we were shoved through standing up or whether we crawled."

All that mattered was that they reached safety—and lived.

John Bradley, the bartender in the Melody Lounge who had sent young Tomaszewski to fix the light, later gave his account of the start of the conflagration:

> The first thing I noticed was that somebody had put a light out in the corner. I told Stanley to go over and put it back on. I went about my work and then I heard somebody say "Fire!" There was a flash that ran across the ceiling.
>
> I jumped from behind the bar. The palm tree was blazing. I pulled it down and tried to throw water on it. I called to the boys to bring me water. I threw two or three pans of water on it, but it was no use. The whole ceiling was ablaze.
>
> People got panicky. I hollered to them to take it easy. Then two or three employees came running in from the kitchen with fire extinguishers, but by that time the smoke was so blinding that it was impossible to do anything. I finally escaped through the kitchen.

A gray-haired waiter named Henry W. Bimler showed considerable coolness and imagination. His first reaction to the fire was to run down into the kitchen and try to get out through the employees' entrance. To his dismay he found the door locked.

Bimler called to the dishwasher for the key, but he proved to be a man who went by the book. "Only the boss can let you have that key," he said.

Baffled, Bimler went back upstairs. He ran into several young women who had been separated from their escorts.

"Can you get us out?" they cried. "Please get us out!"

Bimler guided them back down to the kitchen, which seemed the safer place at the moment. His eyes fell on the nightclub's huge ice box. He and the others entered it and remained in their frigid haven until firemen finally entered the gutted building, broke open the door, and freed them.

Eddie Pierce, a member of the dance team of Pierce and Roland, became a hero during the fire. Pierce, his wife and partner Dorothy Roland, acrobatic dancer Miriam Johnson, violinist Helen Fay, four chorus boys and a young woman were in a small room on the second floor of the attached tenement house, waiting for their cues to go on stage, when smoke billowed up the stairway from the night club. Pierce realized that their only means of escape was the roof twelve feet above them.

He climbed onto the windowsill and, balanced there precariously, hoisted his companions, one by one, to the roof. They then pulled Pierce to safety, and all were later rescued by firemen.

There were other heroic rescue attempts, some of them successful. Band leader Mickey Alpert left the stage at the first indication of fire and ran outside. However, he made repeated trips into the blazing building, dragging people from the fire. He was seriously burned, but survived.

Frank Balsanini, head waiter, also rescued unconscious men and women from the blazing building, but he was not as fortunate as Alpert. Overcome by smoke and badly burned, he finally collapsed on the sidewalk and died.

Perhaps the greatest hero of all was nineteen-year-old Marshall Cook, a chorus boy. As he started downstairs toward the stage for the show, he was driven back by flame and smoke. He smashed a window in the second floor dressing room and assisted thirty-five performers and stage assistants to an adjoining roof. There he found a small ladder. He and a couple other young men held the ladder over the edge of the roof, allowing the others to descend as far as they could. Below, a group of sailors caught those who leaped from the ladder.

Another hero was Joseph Lawrence Lord, a second class

petty officer in the Navy and a former fireman. Lord was driving past the Cocoanut Grove in an automobile when the fire broke out. He leaped from his car, broke a window of the club, and crawled into the fiery interior.

"The smoke was choking and thick," he said, "but no different really than many a fireman meets during many a fire. I crawled along on my hands and knees and then bumped into five forms. All were moaning and some were twisting around on the floor, clawing at their throats. There were three women and two men, and one by one I dragged them to the window I had jumped through. I hoisted them up to the sill and then yelled like hell. Firemen gave me a hand and pulled them through the window to safety."

Lord was puzzled by the fact that more people didn't try to break windows and jump out, but the fact was that most of the windows were concealed by heavy draperies. "The folks I rescued," he said, "looked as if they had tried to sit quietly at their tables but finally passed out and fell to the floor. They must have been struck down quickly by fumes."

Nina Underwood, a Red Cross nurse's aide, told of arriving at the disaster scene and seeing many being treated for hysteria, others wandering the streets in a daze, and many lying dead on the sidewalk. "I had to knock down two survivors with jiu jitsu," she said, "because they got out of hand. I questioned a girl who was pretty high on liquor and she said she didn't know there was any trouble until the smoke grew heavy."

Miss Underwood said she spent some time instructing taxicab drivers on "how to make their overcoats into stretchers and blankets" and telling them that "unless the people were covered up when they got into the air their skin would fall off." In all, she made fifteen trips to hospitals, carrying the injured. "I gave artificial respiration three times during the trips," she said, "and treatment for shock once. In many cases, before getting bodies into the ambulance, we found that rigor mortis had set in when we tried to cross their arms. On the last nine trips we made, all the victims we carried were dead."

The fire did not last long. It quickly worked its way through fragile decorations; consumed tables, chairs, and the wooden bars; and ate through the roof—all within minutes. The Fire Department arrived almost immediately, but still they could do nothing but put out the fire with streams of water and enter the building to carry out the dead—and the few lucky enough to still be alive. It was then that firemen found bodies piled in tangled heaps at the exits, evidence of the terrible struggle to reach safety. "It was hard to believe," one fireman said, "but we found some of the bodies actually pulled apart."

As soon as the news of the fire spread, the city mustered all available equipment and personnel for rescue: more than a hundred ambulances from twenty-two hospitals in the city and suburbs; doctors and nurses from all over the city, as well as outlying areas; naval officers and enlisted men stationed in the vicinity.

Crowds of the curious also flocked to the scene, and, as in all major fires, these often hampered rescue attempts. Seeing that police in the area were outnumbered, several naval officers ordered their men to lock arms and form a living chain to hold back the crowd. Soldiers and Coast Guardsmen helped firemen stretch hoses; priests arrived to administer last rites to victims that lay dying on the sidewalks. When it was found that the ambulances could not handle all of the injured, almost a hundred taxicabs were pressed into service. Trucks moved in to haul away the dead.

The Boston Chapter of the American Red Cross promptly sent 500 trained first aid volunteers under the direction of a dozen professionals. In addition, the New York Red Cross sent six skilled disaster relief workers. Civil Defense units, mobilized for wartime duty, rushed to the aid of the injured and dying.

Confusion was at its height during the first hour following the outbreak of the fire. Some patrons who escaped with only minor injuries went directly to their homes before the police could obtain their names. Two emergency dressing stations were set up in a garage and a drug store to treat minor injuries.

As the injured began to pour into local hospitals, a new threat developed. The supply of sulfa drugs used in burn treatment ran low and the Red Cross put in a hurried call to New York for assistance. Within an hour, a special plane reached Boston with an adequate supply. There was also a shortage of blood plasma until public officials released supplies from emergency blood banks that had been set up by Civilian Defense authorities.

By midnight the area around the burned-out Cocoanut Grove was crowded not only with onlookers but with overwrought people trying to learn the fate of loved ones. The crush of people became so great that at 1:58 A.M. martial law was declared and a mixed force of policemen, soldiers, sailors and Coast Guardsmen formed a solid line to prevent anyone from coming closer than three hundred yards. Hundreds of grief-stricken men and women then besieged the hospital and the city mortuaries in an attempt to locate relatives.

The search for victims or survivors inside the blackened hulk of the once plush Cocoanut Grove continued until daylight. Rescuers found heartbreaking scenes in the charred ruins. The body of a young woman was found in a telephone booth. In her fright, she had apparently attempted to call for aid when she was trapped by the flames. Husbands and wives were found locked in each other's arms at doorways. Piles of tangled corpses, blackened by smoke and flame, were removed from the narrow Melody Lounge stairway and the blocked exits. Inanimate objects also told a sorrowful story—more than three hundred fur coats and evening wraps, ruined by smoke and water, in the cloak room, and an equal number of Army and Navy officers' capes that would never be worn again.

Women's accouterments, such as evening bags, vanity cases, and high-heeled dancing slippers, and fragments of torn evening gowns lay scattered about.

When the dreadful evening was over and a firm accounting of the dead taken, it was found that 491 people had perished in the Cocoanut Grove inferno. There followed, of course, the usual investigation. Although the most publicity was given to the fact that young Stanley Tomaszewski had started the fire by igniting a palm frond while fixing a light, this was only a small part of the truth. The ultimate blame must rest with the Cocoanut Grove nightclub itself and with the City of Boston. The club was a fire trap complete with tinder-box decorations, cheap partitioning, and poor wiring, despite the "conditions good" report that Lieutenant Linney had given it. And the City of Boston, at that time, had few effective fire regulations and failed to enforce those it had.

Robert Moulton, technical secretary of the National Fire Protection Association and secretary of its Committee on Safety to Life, declared that "the tragedy was clearly due to gross violation of several of the fundamental principles of fire safety which have been demonstrated by years of experience in other fires and which should be known to everybody." He went on to point out that, following the Iroquois Theater fire in Chicago in 1903, theaters all over the country were carefully regulated to protect the safety of their patrons, but that no such standards had ever been applied to nightclubs, which, he said, were far more dangerous places. "Nightclubs," he asserted, "are commonly located in old buildings made over for the purpose and practically every known rule of fire safety is violated."

One of the clearest condemnations of the City of Boston was the testimony of James H. Mooney, Commissioner of Buildings. He said that he knew of no legal regulations against the use of highly inflammable decorations in nightclubs and cabarets, and pointed out that Boston fire laws did not require nightclubs to install fireproof sprinkler systems or *even to mark exits with clearly visible signs!*

Probably the only happy story to emerge from the Cocoanut Grove fire involved the Boston College football team. Boston College, rated No. 1 in the nation, had played Holy Cross at Fenway Park that afternoon. Boston College was in line for an invitation to the Sugar Bowl on New Year's Day if it beat Holy Cross—a team that had failed to win a game all season. But the underdog gave Boston College an unbelievable whipping, 55-12, dumping it out of the Sugar Bowl bid and destroying its ranking as the top team in the land.

Boston College players had made arrangements to celebrate their expected victory that night, but they were so heartsick at their humiliating defeat that they cancelled the party and went to sleep in their dorms—an act that saved a lot of young lives.

Of course they had planned to hold the party at the Cocoanut Grove.

11

Death at the Circus (1944)

It was Thursday, July 6, 1944. The day was hot and sultry, the temperature soaring to one hundred degrees in the sun. In a huge vacant lot fronting on Barbour Street in Hartford, Connecticut, the main tent of Ringling Brothers and Barnum & Bailey Circus shimmered in the scorching heat.

More than six thousand perspiring people poured into the main entrance to the Big Top, a mammoth oval-shaped tent five hundred twenty feet long and almost two hundred feet wide, its center arena circled by tiers of seats that would accommodate twice the number of persons expected to attend this matinee performance. The crowd was the usual mixture of the old and the young. But about two-thirds of the customers were small children, accompanied by their mothers or grandparents. There were few young fathers in the audience; the United States was in the midst of fighting World War II.

No one seemed to mind the suffocating heat inside the big tent—leastwise, not the children. They thought only of the spectacular performance they were about to see—the wild animals, the clowns, the aerial acts, all the spine-tingling excitement that made this huge circus the Greatest Show on Earth.

Among the many children at the show was a girl about six years old—a winsome, blue-eyed child with delicate features and light brown hair that tumbled to her shoulders. Her fate that day was to be listed officially in cold statistics at the Hartford Police Department as "Unidentified: age, about six years; race, white; sex, female; height, 3′10″; weight, 40 pounds; eyes, blue; build, moderately developed; head circumference, 20½ inches; hair, shoulder length, blond or light brown, curly."

She would eventually become known to the world as Little Miss 1565.

It was now 2:30 P.M. The show had been on for half an hour. First there had been the colorfully costumed clowns performing their zany acts in the three rings and the sawdust track encircling them. This was followed by Alfred Court's wild animal act. Lions, tigers, jaguars and leopards had been put through their paces to the delight of the children and the amazement of their elders. Now, to the sound of sharply cracking whips and occasional pistol shots, the animals were being funneled into three steel-mesh runways that stretched from the huge steel-barred cage, across the sawdust-covered track, and through an exit to the caged wagons in a corral outside the tent.

The non-stop, continuous entertainment traditionally offered by the circus dictated that the next act be ready to proceed with the show, and the Flying Wallendas had already climbed to their precarious perches high in the vaulted ceiling of the Big Top to begin their death-defying turn on the high wire. For the moment spectators divided their attention between the trainers herding the animals into the runways and the five spotlighted members of the famous high-wire troupe poised for their act. The circus band, led by bandmaster Merle Evans, struck up a gay, noisy tune. In the warm tent, wide-eyed children sipped soft drinks and munched on peanuts, hot dogs and crackerjacks. Some explored fluffs of sticky, pink cotton candy.

At the main entrance of the tent a firewatcher was stationed. When he saw the wild animals being directed into the runways he suddenly became apprehensive. He was afraid an accident might occur if the roustabouts disassembling the runways were careless in their work. Huge jacks near the runways held up the seats, and if they were jarred the seats could collapse. The firewatcher, deciding that such a possibility was more likely than a fire, left his post and hurried toward the runways.

That's when the fire started.

It was a tiny blaze near the main entrance, no bigger than the flame of a match. No one saw it at first, not even other firewatchers posted at strategic spots around the canvas shell of the tent. But a tiny flame was all that was necessary to precipitate a giant holocaust. The feeble flame, still unnoticed, grew to the size of a large rose, then licked insidiously at the side wall of the tent. The inflammable canvas fed it, but it was not until a foot or two of canvas was afire that a woman seated near the entrance gasped, "Fire!" Three ushers—Mike Dare, Kenneth Grinnell, and Paul Runyon—saw it at the same time. They started for the blaze with buckets of water, but by the time they reached it the heat of the flame was such that it scorched their clothes and drove them back.

There was no immediate panic in the audience. All eyes were turned toward the growing fire, but everyone seemed certain that it would be put out promptly. The blaze crept up the canvas wall to a height of six feet, then eight, then ten. It was then that people started to get up, filling the narrow aisles. There was still no rush, for there were a lot of exits. A man kept saying, "Take it easy—take it easy!" and most of the crowd did. It was just a matter of leaving the tent now, waiting until the fire was controlled, and then returning to their seats for the rest of the show.

But what the crowd did not reckon on was the incredible speed with which flame would engulf the Big Top. The crimson blaze raced up the side of the tent, spread across the top with unbelievable ease, and began dropping great patches of burning canvas on the crowd. Then the panic began—a wild, chaotic scramble to escape what was suddenly recognized as a fiery threat to life.

By this time most of the animals had been cleared from the big steel-barred cage at ring-center. But trainer May Kovar was having a little difficulty persuading four leopards to leave the ring. They circled the cage, running past the entrance to the runways several times, refusing to enter. With the fire already in the top of the tent and screaming people dashing for the exits, Miss Kovar knew she had to get the animals out quickly. Above the noise of the crowd she shouted to a helper outside the cage.

"Hose down those cats!"

A man grabbed a hose and turned it on the beasts. Miss Kovar then worked them into the runways and the four animals raced to the corral outside the tent.

High in the top of the tent, and therefore right in the path of the rapidly spreading flames, were the five members of the Wallenda troupe—Joe, Helen, Carl, Herman, and Henrietta. They could do nothing but descend quickly and flee. As Herman later explained, "When the flames hit the roof we saw that we had to get down fast. We slid down the ropes and headed for the performers' exit, but people were so crowded there that we saw we didn't have a chance. So we climbed over the steel cage that lines the exit. That was easy for us—we're performers. But the public couldn't get out that way."

Bandmaster Merle Evans and his twenty-nine musicians

were situated at the opposite end of the tent from the main entrance, where the fire started. In true circus tradition Evans ordered his bandsmen to "keep blasting"—a circus term for playing at top volume. The band responded with heavy brass and an amplified organ in an attempt to calm the audience, but the crowd was now in hasty and disorderly retreat. When Evans realized that the fire was gaining momentum he switched to "Stars and Stripes Forever." This unscheduled march was a signal to circus employees outside the tent that there was trouble and their help was needed—the equivalent of shouting "Hey, Rube!" on a carnival ground. In answer to the call, roustabouts and performers headed for the Big Top.

The audience, aware now that the fire was a serious one, took the logical way out. Those in the lower seats rushed into the center arena and tried to reach other exits. Those in the higher seats, unable to crowd into the aisles and determined to clear a path, threw folding grandstand chairs into the performing area. Some of these chairs struck people already in the arena and, as the bulk of the grandstand crowd surged toward the center of the tent, those in front stumbled over the chairs and piled up those behind. Like crazed animals, people still on their feet clawed their way over those who had fallen.

While this melee was going on, a fickle wind helped the fire spread. At precisely the right moment it blew in from the main entrance, causing the sheet of flame to race even more rapidly across the top of the tent. This fact doomed many who otherwise might have reached the exits. The surging fire overhead turned the crowd into a terrified and senseless mass of struggling humanity on the arena floor.

The greatest obstacle to escape proved to be the three runways leading from the animal cage to the outside. These steel chutes, about four feet high, were positioned at either end of the reserved seat section, so that the audience seated between the runways was literally walled in. Had the fire started five minutes later the runways would have been removed, but as it was there had been no time for the roustabouts to remove them.

The runways thus became a barrier of death. Some people reaching the arena floor managed to scramble over them. But many older people and children, unable to do so, fell before the chutes and those behind tried to crawl over their writhing bodies. At each of the three runways injured and dead piled up in heaps.

One man was on his knees before a runway trying to protect a woman who had fallen. He kept shouting, "Give her air!" but people trampled over her recklessly. A boy attempted to lift his grandmother, who had fallen before a runway, and kept pleading for help from adults. No one paid attention. And, as the people piled up, the flaming canvas overhead began to collapse upon them, turning the mound of human beings into a ghastly funeral pyre.

Great tongues of flame licked down from the burning tent top to pick one person off here, another there, like a sniper selecting victims at random. But most of them were set afire by burning pieces of canvas that fell and blanketed a hundred persons at a time. Some were able to tear away the blazing canvas, beat out their clothes with their bare hands, and escape. Others screamed in agony as they burned to death.

One of the greatest tragedies was the fact that mothers and children were often separated from each other in the crush to reach exits. Crying children looked for mothers or grandparents; some mothers whose children had somehow been left behind, tried to make their way back to them against the surging mob.

Not everyone rushed to ring center. Those high in the bleachers dropped to the ground behind the stands and squeezed beneath the tent or escaped by a side exit. One woman threw down her three children, then jumped and herded them to safety. But most headed for the arena—the only escape route they knew.

The enormous Big Top was held upright by six giant poles. One by one, these began to tumble as the flames swept the length of the tent. With thundering crashes, they fell and crushed people who happened to be in their path.

As the fire ended its orgy of destruction, the circus grounds became a bedlam of noise. Mingling with the shrieks of children and the screams of women was the frightening noise of the animals just outside the tent. Elephants, their tails stiffly extended in fright, trumpeted heavily; lions, tigers and other cats roared threateningly; monkeys jabbered excitedly in their cages. Roustabouts attempted to quiet the animals, fearing that if they became uncontrollable and escaped they would add another dimension of horror to the scene.

One man caught up in the mass exodus from the fire was Thomas E. Murphy, a writer for *The Hartford Courant*. He was attending the circus with his five-year-old son, who had persuaded him, somewhat against his will, to take in the performance despite the stifling heat of the day. As the fire spread, Murphy, with his boy riding on his back, fought his way to the center of the arena. Approaching the runways that blocked the crowd's path, Murphy saw people piling up helplessly before them. "I saw one woman fail to make it over the runway," he recalled. "She slid back and slumped to the ground. A man tried to fend off the crowd but the pressure was too great. I was slammed against the steel barrier and my knee caught momentarily between the bars. Then, taking my five-year-old son in my hands, I tossed him over the barrier to the ground beyond. The flames at this point were nearly overhead and the heat was becoming unbearable."

Murphy managed to scale the barrier, pick up his boy, and race for the exit. When he got there he looked back. People were still engaged in a frenzied struggle to get over the runways as flaming canvas fell on top of them. Outside, Murphy beheld a different, but still heartbreaking, scene: children running around wildly, crying for parents; mothers searching for their youngsters; men and women staring vacantly, too deeply shocked to move or talk.

One terrified woman kept saying, "My four children—where are they?" Then she saw one running toward her, and when she finally managed to round up all four of them she said over and over again, "Thank God! Oh, thank God!"

A policeman had to prevent another distraught woman, her clothing already charred, from fighting her way back into the fire. The tent had collapsed by this time, and the flames were consuming the dead and dying on the ground. The woman fought like a maddened beast, screaming, "Let me go! Let me go! For Christ's sake, my kids are in there!"

Merle Evans and his band, from their position at the far end of the tent, probably had the best overall view. Evans said he watched as the fire crept up the wall at the main entrance, set the tent roof aflame, and made its way toward him. "It just kept coming," he said, "and as it raced the center poles, burned from their grommets, fell one by one." Not until all the poles had fallen and the bandsmen's uniforms were scorched, did the musicians leave.

The fire was a short one. From the time the feeble flame appeared at the main entrance until the entire tent was down in smoldering ruins, it lasted not more than ten minutes. Some insisted it was much less than that.

The Hartford Fire Department acted with great efficiency, arriving within minutes, but they were too late to do anything but hose down the smoldering ruins. They concentrated first on the grisly piles of charred bodies, and then went in to pull out those few who still showed life. When that task was done, they sprayed the rest of the smoldering oval of charred ground.

Within the ten minutes or less that the fire raged, 168 people died and more than 480 were injured, many seriously. Two-thirds of those killed were children and almost all the rest were women. Without doubt, the holocaust was the greatest tragedy in circus history and one of America's worst fire disasters.

The entire city of Hartford heard of the disaster within a few minutes by radio. Soon relatives of those who had attended the circus arrived in cars and on foot. Frantically they stormed police lines, unaware that some of the policemen were worried about their own children, who had attended the performance. One was patrolman James F. Healey. He was on duty outside the tent when the fire began, and he first saw it when it reached the top of the tent.

"It was hardly as big as a cigarette burn," was the way he described it. "But it spread as a cigarette burn does. Then it burst through, suddenly, in a big common fire and went flying around the place."

Healey's wife and his eight-year-old son were inside. Mrs. Healey, seated high in the stands, threw her son into the arms of a roustabout near the wall and then climbed down herself. Both escaped with minor injuries, but Healey was in agony for forty-five minutes before he finally found them.

Help rushed to the scene of disaster in every form—fifteen hundred volunteer workers; one thousand nurses aides and staff assistants from the Hartford Chapter of the American Red Cross; Connecticut State Police; Hartford Police; civil defense units; soldiers from nearby Camp Bradley; nurses and doctors from miles around; priests, to administer the last rites to the dying; and fire apparatus from Hartford, East Hartford, West Hartford and Bloomfield. But because of the speed of the fire, all arrived too late to stem its destruction. Instead, rescue workers busied themselves taking the unidentified dead to the State Armory and the injured to hospitals. More than one hundred ambulances and other vehicles were pressed into service, and when the Municipal Hospital close to the scene of the fire was full, the less seriously injured were transferred to other hospitals.

Hartford—not to mention other cities all across the

country—was sickened by the tragedy. But no one felt worse than the performers and others who made up the community called the circus. With grim faces, they sat for hours after the conflagration, in the shade of circus wagons or outside small tents. Tears rolled down the cheeks of many—for the people, and especially the children, who had perished in the fire, and for the circus itself which had been home to them. Even Gargantua, the great ape, seemed stunned. He stood in his cage, only forty feet from the burned-out tent, gazing solemnly at the smoking ruins.

One group of midgets—a woman and three men—sat in grave silence in the shade of a wagon, staring at the charred outline of the Big Top, the twisted bars of the animal cage, the remains of once-glittering circus paraphernalia that had blackened but not burned. A newspaper reporter approached them.

The man got up wearily and walked away, blinded by tears. All he managed to say was, "We can't talk. We're broken-hearted. We feel sorry for all those people. We can't say anything."

Bandmaster Merle Evans sat on a bench in the sun, staring at the ground. "I have been through storms and blowdowns and circus wrecks," he mumbled, "but never anything like this. I hope to God I never see a thing like this again."

Felix Adler, the internationally famous clown who had been with the circus for thirty-one years, told how he was putting on his makeup in his dressing tent when the fire started.

> We heard a roar like the applause when one of the big acts comes off, only we knew that the animal act was over and there shouldn't be applause. We knew right then that something was wrong. Then we smelled smoke. We then went over to see what we could do to help. I thought the menagerie fire in Cleveland was the worst thing I could ever see [Note: this was a circus fire in 1942 in

> which no human deaths occurred but in which forty animals perished in flames or were shot after breaking out of their cages]. But this is the most terrible. Nobody in the circus business has ever seen anything as horrible as this.

A roustabout probably summed it up as briefly and graphically as anyone. "It was like you'd opened hell's doors," he said.

Immediately after the fire an investigation got underway. Five circus officials—J. A. Haley, vice-president; George W. Smith, general manager; Leonard Aylesworth, boss canvasman; David Blanchfield, chief wagon man; and Edward Versteig, chief circus electrician—were arraigned before the Hartford Police Court on technical charges of manslaughter. The electrician was included because it was thought that the fire might have resulted from a short circuit.

Meantime, serious charges were also leveled against circus management by Hartford County State's Attorney H. M. Alcorn, Jr., after painstaking examination of the charred ruins of the Big Top and extensive interviews with ushers and other circus employees. Alcorn said:

> There appears to have been inadequate fire-fighting equipment on the grounds provided by the circus management and inadequate personnel to operate the small amount of equipment available. Some passageways designed as exits were blocked with animal cages and other equipment. There is also in our possession information that the tent itself had been in use only since the road show started this season and that it had been treated with paraffin which was diluted with gasoline, making the entire tent highly inflammable.

Alcorn's statement about the hazardous treatment of the $60,000 tent canvas with paraffin and gasoline turned out to be fact. Evidently this was a stop-gap arrangement, since the chemicals necessary for effective fireproofing had been designated by military authorities for war purposes. Twenty-four hours after the Hartford fire, however, the government released better fireproofing material to all circuses in the country.

Although the amateurish fireproofing of the tent indisputably aided the rapid spread of the flames, officials were unable to pinpoint how the fire actually started. The electrical shortcircuit first suspected was finally ruled out. It was suggested that, with more than six thousand people in the audience, it was not unlikely that a carelessly discarded cigarette or match could have started the conflagration. This, however, was pure supposition. The true beginning of the fire remained a mystery for six years. Then, in July 1950, the truth came out.

Robert Dale Segee, a twenty-year-old, stoop-shouldered youth, confessed that as a fourteen-year-old boy he had deliberately set fire to the canvas at the main entrance.

To Ohio police, Segee sobbed out a story that was almost unbelievable. At first officers thought he was an unstable, publicity-seeking crackpot making a fake confession. But they were finally able to build a mass of evidence that convinced them that the mystery of the circus fire in Hartford had at last been solved.

Apparently enjoying his opportunity to confess, Segee told police of a lifetime of crime. He related how he had killed a nine-year-old girl with a rock, had strangled three other people, and had set fire to a store, a boat pier, a Salvation Army center, a schoolhouse, and various other buildings—and finally had capped all the rest with the tremendous blaze at Hartford. He claimed he was driven to such acts by a rider on a fiery red horse who came to him in his dreams. Each time the rider appeared, Segee found it necessary to relieve his tensions by putting the torch to something. In Segee's twisted mind the rider was to blame for all his depredations.

It became evident to investigators that Segee's confession gave him a feeling of importance, for it was the first time in his blighted life that anyone had ever paid attention to him. His excuse for his heinous crimes was that "I never got past the second grade, my father never had a steady job. All my life for any little thing I done I was beaten. If you had a bunch of brothers that called you 'dopey' all the time, maybe you'd understand. I never had a happy day in my life."

Segee was sentenced on November 4, 1950 on two counts and received a comparatively mild two to twenty years on each.

What happened to the circus after the fire? The show returned to Sarasota to regroup and decide what to do. One major decision was that the show would never again use a tent, but would appear only in indoor amphitheaters and stadiums or in open air arenas such as ballparks. In time it returned to the road and for seven years played to pay off its debts—676 suits by relatives of those who died in the fire. Not one case went to court and in no instance did the circus attempt to avoid payment. Damages in each case were arbitrated by impartial judges and the circus made the assessed payment. The final total of claims paid was more than $4,000,000, and since the circus carried only $500,000 worth of liability insurance its profits for the next ten years went into paying claimants. The arbitration agreements have gone down as one of the most forthright and honest settlements in modern legal history.

But one strange facet of the hideous fire still remains an intriguing mystery to this day. Hartford police spent many days identifying each charred corpse, and eventually they reduced the number of unidentified to six. Five of these were so badly burned that recognition was impossible, but they were eventually traced to persons who had attended the circus and then been reported missing. The sixth body was that of a young girl about six years old. Her identification should have been easy, since she had died of asphyxiation, untouched by the flames, and should have been easily recognizable. Yet no one among the hundreds who looked at her recognized her and no one ever claimed her.

Detective Tom Barber and his assistant, Edward Lowe, in charge of identifying the victims, found this difficult to understand. Someone must have taken her to the circus.

Someone in her neighborhood—mothers, playmates, neighbors—must have noticed that a little girl was missing. But no one came forward.

The little girl was buried near Hartford in a nonsectarian cemetery cared for by the Hartford Park Department, and the attempt to identify her went on for months. When schools opened in the fall the two detectives asked teachers if a little girl was missing from their rosters. They talked with mailmen, tradesmen. They contacted local churches, welfare organizations. They urged newspapers and magazines to publish her picture, taken as she lay on a slab at the morgue. They distributed a dental chart to thousands of dentists. Radio broadcasts, newspaper ads, and widely distributed circulars were used. Missing persons lists were assiduously scanned. And when letters poured in from all over the country from people trying to help, Barber and Lowe tracked down each lead with painstaking care.

Nothing worked.

Completely baffled and frustrated, the Police Department decided that the least they could do was to place flowers on the little girl's grave on Christmas, Memorial Day and July 6, the anniversary of the fire. For years the two detectives decorated the grave.

On July 6, 1974—the thirtieth anniversary of the horrible disaster at Hartford—a seventy-eight-year-old man visited the grave of the little girl and placed a bouquet of flowers near the headstone. He was detective Tom Barber. His partner, Lowe, had died eight years before.

Barber, retired in 1962, still considers identification of the girl a personal mission. Even now an occasional clue will drift into his hands, and when this happens he traces it down. But the child victim of the Ringling Brothers and Barnum & Bailey fire is still unknown. The only identification on the small gravestone is the morgue number assigned to her.

It says: Little Miss 1565.

12

America's Worst Hotel Fire (1946)

WHEN the Japanese attacked Pearl Harbor on December 7, 1941, President Franklin Delano Roosevelt described it as a date that would "live in infamy." Five years later—on December 7, 1946—a surprising and shocking event brought a personal day of infamy to the people of Atlanta, Georgia. It was the day the "fireproof" Winecoff Hotel caught fire.

W. Frank Winecoff had every reason to be proud of himself and his new hotel. The year was 1913 and Winecoff was only thirty-seven years old—a tender age, he thought, to be the owner of a major hostelry on Peachtree Street in the heart of Atlanta's business district. Winecoff had no doubt that his sortie into the hotel business would be a financial success. Atlanta was a growing city, and both casual travelers and hurried businessmen were constantly looking for good accommodations. Peachtree Street was an ideal location for capturing this itinerant trade.

Winecoff was aware that his place lacked the ornate elegance of larger and more expensive hotels, but the Winecoff was a substantial fifteen-story building with 210 comfortable rooms. Besides, it had the advantage of being built exclusively of non-combustible materials. The walls and floors were of steel, reinforced concrete, face brick, marble and terra cotta. The dividing walls were of hollow tile. This superior construction, in Winecoff's mind, made his hotel one of the safest in Atlanta, and he lost no time in advertising it as "fireproof." It was a statement few, if any, other Atlanta hotels could make.

As expected, the new hotel prospered and Winecoff ran the establishment for twenty-one lucrative years. Then, in 1934 at age fifty-eight, he retired, leasing it to a hotel man from Birmingham, Alabama. After that the hotel changed hands a couple of times, and in the fatal year of 1946 it was under the management of Arthur Geele, Sr., and Robert O'Connell, both of Chicago. Through the years, however, Winecoff arranged with each new management to occupy a suite on the tenth floor as a life-time, nonpaying guest.

Even by 1946 the Winecoff Hotel had changed little. It was a squarish building, small in ground area, with its fifteen stories rising in tower-like fashion. On each floor, except for the ground-level lobby, were fifteen rooms. Six of them faced Peachtree Street, with another six on the alley that ran parallel to Peachtree, two on one side street, and, on the other, a room that could be used either for sleeping or for sales presentations. In the center of the building were two elevator shafts and a stairway that circled the elevators. Since the hotel was regarded as fireproof, there were no outside fire escapes, no sprinkler system, and no automatic fire alarm.

The fire that brought a tragic end to the Winecoff Hotel started while most of the 285 guests were asleep. It was Saturday; the time approximately 3:30 A.M.

There was little activity at that time of the morning.

Comer Rowan, the night clerk, was drowsy and bored. He manned the registration desk and operated the telephone switchboard, but this double job made no great demands on him. He amused himself by looking out the front doors at deserted Peachtree Street, now bathed in the pale light of a December moon. Only an occasional passing automobile enlivened the scene.

Rowan's night of boredom was about to erupt in more excitement than he wanted. For at that moment, on the third floor, a mattress left temporarily in a corridor was about to burst into flame. It had been smoldering for some time and now tiny tongues of fire were beginning to emerge.

At 3:32 a light on the switchboard caught Rowan's attention. It was room 510. A man's husky voice said, "I'd like you to send up some ice and ginger ale."

"Right away, sir," Rowan said.

The night bellhop, Bill Mobley, was in the lobby, talking to the hotel's building engineer.

"Take some ice and ginger ale to 510," Rowan said to Mobley. Then he turned to the engineer. "Why don't you go along and check out the building?" he suggested.

The engineer nodded. A night check of the premises was a routine task, and it was about time for it. He waited until Mobley got the ginger ale and ice and the two men took one of the two elevators to the fifth floor. There they separated, leaving the elevator door open so that it would be available for their descent.

As Mobley approached 510 he could plainly hear a shower running. He knocked gently on the door.

"Be right out," said the man inside.

Mobley waited about three minutes before a half-clad man opened the door. He went in, placed the ginger ale and ice on a small table, accepted a tip, and started to leave. But when he opened the door again he found the hall gray with smoke.

Meantime a woman elevator operator was taking a guest to the tenth floor in the other elevator. As she came back down she smelled smoke coming from somewhere between the third and fifth floor. Frightened, she reported it immediately to Rowan at the desk.

Rowan acted swiftly. "Go back up to the fifth floor and find the engineer," he ordered. As the girl returned to the elevator, Rowan ran to the stairwell in the lobby and looked up. To his amazement he saw the reflection of flames a few floors above his head.

Rowan raced to the telephone and called the fire department. It was precisely 3:42. Then he began phoning the rooms, alerting the guests to the emergency. He had completed only a few calls when the phone went dead. The rest of the two hundred eighty-five guests received no warning at all.

Meantime, the fire in the abandoned mattress leaped up newly painted walls, raced down the corridor to the stairwell, and then, whipped into a frenzy by a strong draft coming up the stairs, became a giant tongue of flame reaching for the higher floors.

At the moment flames raged up the stairwell, the guests were automatically doomed. No one—Winecoff or any of the other managers—had recognized (or perhaps they had ignored) the fact that if a fire started in the center of the square building, immobilizing the two elevators and rendering the stairway impassable, there was absolutely no safe way out. Lacking outside fire escapes or a sprinkler system, the guests were trapped.

The Atlanta Fire Department responded to Rowan's call with amazing speed. Within little more than a minute the first engine was on the scene. Within ten minutes the city's full 60-piece force had arrived.

But they were too late. The Winecoff Hotel was ablaze from the third floor up.

Fed by the draft from below, flames and noxious gases

skyrocketed up the two elevator shafts and the surrounding stairway. Like a monstrous blowtorch the fire roared, spit and sizzled its way to the fifteenth floor, effectively cutting off escape for those above the third floor. Balked by a ventless roof, the flame doubled back on itself, seeking other outlets for its fury. The only outlets were the hallways on each floor.

Shooting down the hallways like flame throwers, the flames licked at closed doors, melted doorknobs, turned plaster walls into powder, leaped through transoms to devour people sleeping in their beds. If the fire itself did not reach them, the smoke did, seeping into rooms and killing people by asphyxiation. Some of the guests, alerted by a call from Rowan or awakened by the sudden heat, made the mistake of opening their doors; the fire rushed inside, turning them into fiery pillars. Others, aware that there was no escape via the hallways, went to their windows. But one simply does not jump from fifteen floors above the sidewalk, nor ten, nor even five. And yet, they had to get out some way, for the rooms were filling with death-dealing smoke.

As a frantic last resort, people began tying bedsheets together, fastening them firmly to the legs of beds or other heavy furniture and dangling them from the windows. The hope was that they would be able to hang outside the smoke-filled rooms and perhaps survive until firemen could bring them down.

When the firemen arrived, the building was burning like an open hearth furnace. Bedsheet ropes were dangling from many windows, some with people hanging from them. Other guests had climbed onto ledges to await rescue. Those unwilling to take these risks stood at open windows, screaming for help. The firemen had two choices. They could spray the fire with hoses, or they could let the fire burn and concentrate on rescue work. The first choice appeared to be impractical because the fire had already reached a stage where it would take hours to subdue the blaze, and in that time many would die. The second choice, rescuing as many people as possible quickly, seemed more promising.

With doggedness and courage the firemen ran up their spindly ladders and attempted to pluck people out of windows, off ledges, or from hastily fashioned bedsheet ropes. In some instances they succeeded, but in others people clinging to these makeshift ropes lost their grip, knots came loose, or the supports for the ropes were burned away, and the victims went crashing to the sidewalk.

Then the inevitable happened. Terror replaced common sense and the people began to jump.

To those watching from the street, the sight of fear-crazed humans leaping from windows was the most horrifying sight of all. These were people who had been forced to their windowsills by either dense smoke or flames at their backs. It seemed not to matter any more what floor they were on—they simply jumped. It was the only avenue open to them, and rather than burn in agony, they took the chance.

The firemen, realizing their intentions, begged them not to jump, but the frantic guests were now mass-bent on suicide. One by one they left their precarious perches, screaming and wailing as they fell, their voices cut off sharply as they struck the sidewalk and died. The firmen hastily spread a safety net and caught two girls that jumped from the tenth floor. But they caught very few, because there were so many.

One man, clinging desperately to a bedsheet, was trying to lower himself toward one of the ladders when two persons leaping from several floors above struck him as they fell. All three crashed to the sidewalk below. A woman jumped from one of the upper floors and struck a steel cable supporting the hotel's marquee, her body hanging there by the neck for a moment before plunging to the street. Another woman came to her window on the seventh floor, her agonized face silhouetted against a background of pink flame. First she threw her small son out of the window, followed by a younger girl. Then she jumped herself. All were killed. In another case a woman leaped from an upper floor and struck a fireman carrying a woman down a ladder. All three were tossed violently to the sidewalk. Both women were killed and the fireman seriously injured. A man missed a life net by inches, ripping the coat off a fireman in a last effort to save himself.

Others, afraid to jump, succumbed to smoke and flame. Three shouting persons appeared in windows on the thirteenth floor. Suddenly all three were enveloped in angry red flames and disappeared from sight.

The jumping craze seemed to spread. What one attempted unsuccessfully, another seemed eager to try. Body after body hurtled down, hitting the pavement with thick, leaden sounds. Peachtree Street and the alley parallel to it became littered with dead. But yet more tried it. A girl threw a makeshift rope of bedsheets from her window and attempted to lower herself to a firemen's ladder. With her feet braced against the scorched walls of the hotel, she made her way slowly down the rope. Suddenly her feet slipped and she spun around crazily and let go. She struck the marquee and was killed. Another woman leaped from a seventh floor window toward a net held by firemen. As she fell she screamed, "I hope I live! I hope I live!" She did, but she suffered a broken leg, arm, and hip in the fall.

Some were lucky; some were not. One man threw out a bedsheet rope, stepped back, and never reappeared. He had been overcome by smoke. A girl in a thin nightgown stood like a statue on her windowsill, apparently summoning the nerve to jump. As she stood motionless flames licked out and set the nightgown ablaze. She leaped, a flaming torch.

Many people stayed at their windows, praying or screaming, until flames claimed them. One man stood at a window that glowed with fire, his head rolling from side to side, until he sank below the windowsill in death. On the thirteenth floor a group of people pooled their bedsheets to make an extra long rope. Then, in ill-considered haste, two of them started to descend at the same time. The sheets parted and they fell to their deaths. At almost the same time a four-year-old boy was hurled from an upper window and would surely have been killed had it not been for a man on the street who caught him.

An army major staying at the hotel showed bravery and imagination in rescuing his aged mother. Major Jake Cahill and his wife waited in their room until a firemen's ladder reached their window. Once he and his wife were safely taken down, Major Cahill raced around to the alley beneath his mother's room. He entered the adjacent building and placed a plank across the alley to his mother's windowsill. He crawled along the plank, urged his mother from the window, and guided her safely across his makeshift bridge. While he was performing this act of mercy, someone stole his fountain pen and travelers' checks from a coat he left behind.

Twenty-one-year-old Charles Boschons and his bride, Mildred, were honeymooning at the Winecoff on the night of the fire. "We were on the twelfth floor," the young bridegroom said. "Loud clanging caused me to awaken. It must have been four o'clock. Smoke filled our room and I could see flames outside the window."

Boschons opened the door but immediately had to close it because blinding smoke entered the room. Realizing there was only one way to get out, he tied several bedsheets together and fastened the end of the flimsy rope to a radiator. "There was a fire ladder reaching to the tenth floor," he related afterward. "My wife started down the sheets, hand over hand, the way I told her. She got to the eleventh floor, but then lost her grip and disappeared. I saw her fall. I don't know how I was able to make it after that, but I did, and the fireman grabbed me."

Another young man the same age as Boschons suffered a similar tragedy. He was G. D. Burch who had come from Chattanooga with his wife, Edith, on a vacation. "I was on the tenth floor when screaming woke me up," he recounted. "The exits were blocked off with smoke. I knotted several sheets together. I've been in the army and had some experience in rope climbing. The firemen's ladder reached up to the eighth floor. My wife went out first, but slipped."

Stunned by the sight of his wife falling, Burch managed to climb down and was taken to nearby Grady Hospital in a state of hysteria. He kept calling for his wife until an intern overheard him.

"Your wife is alive," he said. "We've been treating her."

Burch's relief, however, was short-lived. Two days later his wife died. She had survived a fall of ten floors and had astounded doctors by holding onto life for forty-eight hours.

Two women on the eighth floor managed a miraculous escape. They were twenty-four-year-old Ima Del Ingram and her aunt, Mrs. W. R. Tribble. When smoke poured into their room, Miss Ingram knotted sheets and lowered them from her window to a firemen's ladder two floors below. Mrs. Tribble went first, but could not hold on and hurtled into the black of the night. Shaken, Miss Ingram reached the ladder safely. Suffering from shock, she was taken to a hospital and was surprised to find that her aunt was also there.

"I lost conciousness as soon as I slipped," Mrs. Tribble said. "The next thing I knew I was in the hospital. I guess I must have hit a net."

Luckiest of all, perhaps, was Franklin S. Fader, a businessman from Brooklyn. He left the hotel three minutes before the fire broke out to meet friends at the airport.

There were many incidents that pulled at the heartstrings. One man, separated from his two children, spent hours looking for them in hospitals. He finally found them—dead and burned almost beyond recognition. An eighteen-year-old girl, rescued from her eleventh floor room, kept pacing back and forth in front of the hotel crying, "Where is my sweetheart? Where is my sweetheart?"

A sixteen-year-old high school girl, Dorothy Moen, was in the hotel attending a State YMCA Youth Congress. When her room filled with smoke she leaped from her window, crashing to the pavement five floors below. She suffered several broken bones but survived. Eighteen of the 51 young people attending the YMCA meeting died.

Major General P. W. Beade and his wife were awakened at 3:30 A.M. by cries for help. Finding the corridor clouded with heavy smoke, Beade closed and braced the door and then he and his wife went to the window to calmly await help. They were on the sixth floor and were rescued by a firemen's ladder.

"I spent ten months on active duty with the famous Thirty-Fifth Division during the war," Beade said, "but this fire was worse than anything over there. At least you felt you had a chance dodging bullets, but you're just helpless when you're trapped in a hotel room with roaring flames all around you."

One of the most hair-raising escapes from the holocaust was that of Mr. and Mrs. Robert Bault. They were awakened by flames creeping into their fourteenth floor room. Like others, Bault decided to use the bedsheet to escape, but instead of dangling it from the window he stretched it along a narrow cornice from his room to that of his neighbors, Mr. and Mrs. J. B. Phillips.

Phillips' room was a doubtful sanctuary, but nevertheless Bault and his wife crawled perilously along the cornice while people gathering on Peachtree Street watched in horror. When the Baults entered Phillips' room they found he had kept the flames out by stacking the mattresses against the door and wetting them down with water from his bathroom. All four were rescued by firemen.

Mrs. W. Y. Johnson was in a ninth floor room when she was awakened by screaming and the penetrating odor of smoke. She opened her door, then slammed it shut as smoke invaded the room. Her bedroom window would not budge. "The small bathroom window was the only one I could open," she said later, "and firemen finally ran a ladder to the window and I managed to wiggle through to safety."

Eventually firemen were able to enter the lobby of the hotel with hoses and make their way up the stairway against intense heat and clouds of smoke. It was a long, hazardous four-hour job, fighting their way step-by-step, floor by floor, playing their hoses on the flames.

As they entered room after room, looking for the dead or the living, the firemen came upon pitiful sights. Some victims were burned to a crisp by rampaging flames; others, untouched by fire, had died of asphyxiation. In one room on the eleventh floor they found a mother kneeling in the bathroom holding three children in her arms. They had been fused by the fire. In another room they found an entire family dead of smoke inhalation. The father lay with his head in the shower, while his wife was sprawled on the floor with her arm around their two children.

In many rooms unburned bodies lay as if sleeping. One woman sat with her head peacefully on a windowsill, still clutching a purse in her hand. In another room four teenage girls were sprawled, their faces contorted in the horror that must have gripped them as they died from inhaling heated air and gases.

On most floors the rooms were gutted. Furniture was blackened, wash basins were cracked from the intense heat, even telephones had been melted into blackish lumps. Many toilet bowls, the water in them converted to steam, had exploded.

The tragedy was compounded for the firemen by finding personal possessions: a box of children's playthings scattered across a charred floor; a pair of nylons hanging over a dresser; whiskey bottles turned black; postcards meant to be mailed the following morning, many bearing the familiar message, "Having a wonderful time."

They found things, too, that defied explanation. In one room was an uncooked turkey in a pan, apparently ready for some kind of party. Even more amazing, surprised firemen found a caged canary in a burnt-out room, singing lustily.

One of the most poignant but humorous stories to emerge from the holocaust was that of Mrs. Banks Whiteman, an elderly permanent resident of the Winecoff Hotel. Her room was on the fifteenth floor—too high for her to descend by a bedsheet rope or attempt a jump. Fortunately her room escaped the brunt of the fire, and Mrs. Whiteman saved her life by stuffing wet towels around the door, keeping out most of the smoke. It was not until 7:45 in the morning—four and a quarter hours from the time the fire started—that she was rescued. Since she was too old to descend the stairs by herself, she was carried by two fireman from the top floor to ground-level over charred steps slippery with water. When one of her rescuers slipped on the wet surface, Mrs. Whiteman's sense of humor came to the fore.

"You'd better be careful, young man," she said, "or I'll have to help you down."

By the time the firemen entered the hotel, there were few residents still alive. But hospitals in Atlanta and adjacent areas, including those at Fort McPherson and the Atlanta Naval Air Station, had been busy for hours treating the injured. Shuttling ambulances carried both the hurt and the dead. Hundreds of doctors and nurses were on the scene, and the American Red Cross had set up emergency headquarters near the hotel to render first aid. Sorrowing families and friends arrived, then drifted to hospitals and mortuaries.

Immediately investigations on the city, county and state levels were ordered. Appalled by the disaster, Georgia Governor Ellis Arnold said angrily, "The public is being defrauded when a hotel is advertised as fireproof but really isn't. Responsible agencies should prohibit the use of the word fireproof when a hotel is really not fireproof—as the Winecoff obviously was not."

Atlanta's mayor, William B. Hartsfield, along with Fire Chief C. C. Styron and Fire Marshall Harry Phillips, launched an investigation that included close questioning of all 125 employees of the hotel. Phillips made a statement that surprised and outraged the citizens of Atlanta. He maintained that the Winecoff Hotel had been inspected a short time before the conflagration and had measured up to the Atlanta Fire Department's safety requirements. This comment implied that the city's safety standards for hotels either were not very high or were not effectively enforced.

Despite the city's apparent culpability, several laxities on the part of hotel management were cited during the investigations.

First, the most glaring oversight was the fact that the hotel was not equipped with outside fire escapes. Total reliance in case of emergency was placed on the two elevators and the stairway. When the hotel was built, knowledge of fire resistance was in its infancy. But the condition had not been corrected as fire prevention techniques improved. (It was later learned that other Atlanta hotels also were without adequate fire escapes.)

Second, the doors leading from the halls to the staircase were inadequate. Since the fire started in a third floor hallway, the flames would have been prevented from getting into the stairway and spreading to other floors had proper metal firedoors been installed.

Third, had similar corridor-type doors been installed in all rooms, the casualties would have been greatly reduced.

Fourth, automatic sprinklers would have extinguished the small mattress fire before it got out of control. A minimum sprinkler system in a hotel covers halls, corridors, stairways and storerooms. A maximum system also covers all rooms. Either type would have helped.

Fifth, an automatic fire alarm system devised to transmit

an alarm to the fire department would probably have been triggered early enough for firemen to quench the blaze before it got out of hand.

Although some of these fire preventatives were unknown when the hotel was built in 1913, they were available in 1946. The obvious conclusion is that the Winecoff Hotel management had simply failed to modernize, and that the city of Atlanta had not forced it to act. (It was found that other Atlanta hotels also were remiss in one or more of the fire protection systems then in use.)

Despite these incriminating findings, charges of involuntary manslaughter against the hotel lessees, A. F. Geele, Sr., A. F. Geele, Jr., and R. E. O'Connell, were dropped six months later.

The Winecoff tragedy was the most disastrous hotel fire in American history. Of 285 guests, 119 died and more than 100 were injured. The casualty toll was much higher than two other noted hotel fires—the burning of the Newhall House in Milwaukee in 1883, with the loss of seventy-one lives, and the 1946 Hotel LaSalle fire in Chicago, which claimed sixty-one lives.

But probably the most ironic touch to the story of the Winecoff Hotel fire was the fact that when rescuers moved through the hotel looking for victims, they discovered the body of seventy-year-old W. Frank Winecoff in his suite on the tenth floor. The man who had built the Winecoff in 1913 had died in the terrible holocaust that devastated his "fireproof" hotel.

13

Terror at Texas City (1947)

THE morning was cool and brisk, with a freshness that comes only from the sea. An off-shore wind whipped the waters of Galveston Bay, on the Gulf coast of Texas, into a frothy pattern. Overhead an awakening sun gradually warmed the cool air.

Texas City—a bustling town born in World War I and grown to maturity during World War II—was ready to face a new day. Its busy landlocked harbor was beginning to stir. Longshoremen and warehousemen were reporting to work on the day shift, which began at eight o'clock. Those leaving the night shift were wending their way toward the parking lots not far from dockside.

It was April 16, 1947.

Texas City had come a long way since the days of the Second World War. Its population had soared from five thousand to seventeen thousand five hundred and what had been a somnolent town had become a thriving and important

Gulf port. Its rapid growth, however, had resulted in a housing shortage as home builders vainly tried to keep pace with the booming population. Athough there were many fine middle-class homes in Texas City, there were not enough low-cost dwellings for the unskilled laborers that worked at the port facilities. These people—many of them black and Mexican workers—lived on the flatlands just behind the dock area in flimsy, overcrowded shacks.

But the town was without doubt an industrial focal point. The government had favored Texas City with the biggest tin smelter in the world and the only one in the country. At dockside the Monsanto Chemical Company had erected a $19,000,000 plant to make styrene, an essential ingredient in the manufacture of synthetic rubber. Next to Monsanto was the Texas City Terminal Railway Company with its huge warehouses, its tangle of tracks and its engines. Nearby were oil refineries with catalytic cracking plants that converted crude oil into high-octane gasoline. At the wharves there were always ships and hustling longshoremen and stevedores loading and unloading cargo.

On this particular morning there were three freighters at the docks, all being loaded. At Pier 0 of the Terminal Railway Company lay the 7,176-ton French line ship *Grandcamp*. She was partly loaded with oil drilling equipment, large balls of sisal twine, and ammonium nitrate fertilizer. At Piers A and B, about six hundred feet from the *Grandcamp*, were two American freighters, the *Wilson B. Keene* and the *High Flyer*. The *Wilson B. Keene* was about to be loaded with a harmless cargo of flour. The *High Flyer* already had nine hundred sixty-one tons of ammonium nitrate fertilizer aboard and was loading 2,000 tons of sulphur.

No one seemed much concerned about ammonium nitrate fertilizer. It was considered a stable chemical that could be jarred and jostled without danger. It *could* explode, if subjected to extreme changes in temperature, but this was considered a remote possibility.

Certainly no one in Texas City could have possibly imagined what was in store for them this day. It was a typical, routine morning. The men of the town had left for work, either at the docks or in the business district; wives were at home; children were either on their way to school or had already arrived. Everything, even the weather, was normal.

But aboard the freighter *Grandcamp* the first intimation of danger was making itself known. In the ship's No. 2 hold were 1,400 tons of ammonium nitrate fertilizer, and in her No. 4 hold, 880 tons. On the dock were piles of 100-pound bags still to be loaded. L. D. Boswell, a longshoreman, stood alongside the open hatch of hold No. 4. Glancing down, he thought he saw a wisp of smoke coming from the hold. Alarmed, he peered down into the cavernous innards of the ship. The odor of smoke assailed his nostrils. Without hesitation, Boswell notified the ship's captain.

Captain Charles de Guillebon reacted with caution. He was reluctant to destroy a valuable cargo by fighting what seemed to be an insignificant blaze with sea water. Instead he ordered the hatch closed. He planned to turn live steam into the hold. The theory was that live steam, while not as destructive as water, would supplant the oxygen in the hold and smother the fire—a technique that often worked. It was 8:20 when the order was given.

Boswell got the hatch closed at 8:30, but within minutes it became evident that live steam would not do the job. The fire was making headway, Captain Guillebon sounded the fire alarm at 8:33 and ordered his crew and all longshoremen to leave the ship.

The minutes ticked by as the fire in the hull of the *Grandcamp* spread. There was no longer any concern about saving the cargo. Fireboats moved in and pumped white plumes of water into the ship. The Texas City Fire Department arrived within minutes. Workers on the dock stopped to watch the operation, unaware of mounting danger. Billowing waves of smoke attracted sightseers to the edge of the dock area. Some workers, less curious, went about their jobs or stayed at their desks.

Henry J. (Mike) Mikeska, president of the Texas City Terminal Railway, arrived just as the fire started and rushed to Pier 0 to help fight the blaze. Henry Baumgartner, a purchasing agent for the company and also a voluntary fire chief, left his desk to go to dockside. Robert Morris, assistant plant manager of the sprawling Monsanto Chemical Company, went outside to watch the fire from the Monsanto pier. T. A. Thompson, chief works accountant at Monsanto, was at his desk, deeply involved with a column of figures. Frank Taylor, employed in a dockside warehouse, was looking forward to a normal working day.

On board the American freighter *Wilson B. Keene*, chief mate Franklin Woodyard watched the growing fire on the *Grandcamp* with some apprehension. He rang the chief engineer.

"There's a fire on the French ship," he said. "How long will it take to build steam if we have to move out?"

"At least an hour," the engineer replied.

Woodyard's eyes turned to scan the *Grandcamp* again. There was plenty of water being sprayed on the ship. Probably they would extinguish the blaze quickly. And likely the *Wilson B. Keene* would not have to be moved.

But by 9:00 A.M. those who knew something about fires aboard ship were thoroughly alarmed. Henry Baumgartner was one of them; Mike Mikeska was another. They were worried that the blaze might spread to other ships and eventually set the entire waterfront afire. Robert Morris, the Monsanto man, knew the characteristics of ammonium nitrate and was more concerned about the possibility of an explosion.

He alerted the company's fire-fighting crew to stand by. Then he commandeered a jeep and started toward the burning ship.

At 9:10 A.M. the scene had not changed. Black smoke rose from the *Grandcamp*, fireboats arched water into the ship, the entire Texas Fire Department was on the dock hosing down the big freighter—and the crowd on the wharves, unaware of the danger, watched the show.

It is estimated that, by 9:12 A.M., nearly four thousand spectators and dock employees were at or near the wharves.

That's when the *Grandcamp* blew up.

The violent blast was one of the biggest in mankind's history. It was not as mighty as the atom bomb blasts that shattered Hiroshima and Nagasaki in 1945, but it was comparable to the terrible explosion of the munitions ship *Mont Blanc* in Halifax Harbor in 1917. Later it was estimated to be equivalent to 250 five-ton World War II blockbuster bombs detonating at the same time. And it started a chain reaction of explosions and fires that all but destroyed Texas City.

The destruction of the entire waterfront was measured in minutes. The thunderous explosion blew the hatch covers of the *Grandcamp* high into the sky where they seemed to ride, for a moment, on a towering cloud of brownish smoke. Then the entire ship simply disintegrated, with pieces of it flying in every direction. Chunks of the superstructure, some small and others weighing tons, soared into the sky like rockets. Huge fragments of the ship's plate were hurled for four or five miles. Many of these red hot pieces of metal headed like well-aimed arrows toward some fifty oil tanks that sat in quiet vulnerability along two miles of the waterfront. When the fiery chunks of steel struck the oil tanks, they burst into flames. Some, hit by larger pieces, crumpled like a tin can under the foot of a small boy.

Pier 0, the dock at which the *Grandcamp* had been laying seconds before, disappeared. The fierce blast gave birth to a giant wave that washed over the entire dock area, inundating cars in the parking lot and drowning people fleeing the disaster. A 150-foot steel barge was blown out of the water and pitched 200 feet, landing in the parking lot and crushing cars under its weight. A fire engine on the dock was picked up and tossed high into the air like a toy. Two light planes, hovering 1,500 feet over the bay, plunged into the water—either felled by the concussion or hit by flying debris. The other two freighters—the *High Flyer* and the *Wilson B. Keene*—were ripped loose from their moorings and jammed together like Siamese twins.

Within one minute of the explosion the Monsanto Chemical Company's styrene plant ceased to exist. Fiery debris from the *Grandcamp* rained down on the huge building, setting it afire. Three horrendous explosions followed at once from chemicals set off by the blazing metal from the ship. The entire plant blew up, collapsing in a heap of smoldering rubble.

But that wasn't all. Terminal Railway Company buildings also collapsed from the force of the blast. A steel and brick warehouse was flattened. Other buildings were crushed or set afire. Pipelines, carrying such inflammable chemicals as benzol, propane, and ethyl benzene, burst. Huge thirty-foot long oil drilling stems weighing over a ton each—part of the *Grandcamp's* cargo—were hurled like spears for 13,000 feet and created havoc wherever they landed. Even the balls of sisal twine carried by the French ship added to the disaster. Converted into flaming spheroids by the explosion, they were catapulted over the entire city, setting fires everywhere they landed. Some of the twine flaked off and descended on buildings and people like ignited confetti.

The instant destruction of human life was almost unbelievable. Within seconds 145 Monsanto employees, 27 members of the Texas City Fire Department, and 82 laborers and sightseers on the dock were killed. Several hundred others were seriously injured.

People died in many ways. Some, fleeing the scene, were strangled by poisonous gases set free by the destruction of the Monsanto plant. Others were cremated where they fell. A great number were crushed in collapsing buildings. Still others—perhaps the luckier ones—were seen fleeing the horror with blood gushing from ears and noses, the result of concussion.

The blast rocked the region for miles around. In Galveston, eleven miles across the bay, windows were shattered and ceilings collapsed. At Port Arthur and Orange, one hundred miles away, buildings shuttered. At Pelly, twenty-seven miles distant, a man complained that "the sound hurt my eardrums." And two miles away a couple sitting in an automobile were found dead. A piece of flying sheet metal had decapitated them.

All through the city instant damage was inflicted on homes and buildings. Walls caved in; windows were pulverized. Children in schools were cut by jagged pieces of glass; flying metal tore gaping holes in buildings. Panic-stricken people took to the streets. They stood in frightened groups, dazed and uncertain, discussing what should be done, not knowing what had caused this sudden calamity.

Within an hour of the giant explosion a reporter from the *Houston Chronicle* flew over the stricken city. He painted a graphic word-picture of the scene:

> Flying toward the blaze, the smoke could be seen from Ellington Field, approximately thirty miles away. It reached four thousand feet. One oil tank, about a thousand feet away from the blaze, was crumpled like a piece of tinfoil. Buildings along the railroad tracks had the ends blown out—the sides were intact. An industrial section close to the bay was afire with the biggest blaze. A refinery near the tracks was also burning. Several oil tanks were blazing brightly. A few tanks were crumpled by the blast. A heavy cloud of smoke hung over the scene, shot with flashes of flame from the fires that still raged on the waterfront.

Hal Boyle, an Associated Press correspondent who also flew over Texas City, said, "In four years of war coverage I have seen no concentrated devastation so utter, except at Nagasaki, Japan, victim of the second atom bomb, as presented today by flaming Texas City."

General Jonathan M. Wainwright, hero of Bataan, agreed with Boyle's description. "I have never seen a greater tragedy in all my experience," he said. "I am here to offer this stricken city every facility that the Army can place at its disposal."

Many individuals who miraculously escaped death described their experiences. One was Ben Lapham, second mate on the *High Flyer*, moored about six hundred feet from the *Grandcamp* when it blew up. He was on deck, watching firemen and the fireboats pump streams of water into the freighter to extinguish the fire. When the *Grandcamp* exploded, Lapham was thrown to the deck.

"Everybody on our ship was standing around watching the fire." he related afterward. "The fire burned about half an hour and then there was the most terrible blast. The funny thing was that everything blacked out. I was knocked flat to the deck. It was like night for a minute or more. The blast blew our hatches open and killed several of our crew. Wreckage rained down for several minutes, but all I got were some bad scratches."

Robert Morris, the Monsanto assistant plant manager who had commandeered a jeep to take him closer to the burning ship, was rolling along the Monsanto pier when the *Grandcamp* blew up. Morris and the jeep were lifted high in the air. When Morris fell to the ground the huge wave from the bay swept over him. The wave dealt death to many, but it prevented Morris from being burnt to a cinder.

W. H. Sandburg, vice-president of the Texas City Terminal Railway, was among the lucky ones. He left the *Grandcamp* five minutes before the explosion. "The concussion was simply terrible," he said later. "It blew out windows of every home in town. It blew in ceilings and destroyed business buildings. It cracked many new buildings from end to end. Doors were blown from their hinges."

In the business district, some distance away from the waterfront, a state highway patrolman walked into City Hall. Just then he heard a deafening explosion. He thought it was a bomb. Looking out of the window, he was amazed to see birds dropping to the ground. "They were dropping out of the air, one after another," he said, "killed by the concussion." Outside he saw the streets littered with broken glass and other debris.

Frank Taylor, beginning his daily work in a dockside warehouse, heard the horrendous explosion an instant before the building caved in. Injured and bleeding, he dug his way out of the wreckage and crawled through a hole in the wall. With a splash, he fell into Galveston Bay. Suffering from shock, the intrepid Taylor swam a half mile to his home on the bay. He found his house reduced to tangled wreckage.

T. A. Thompson, the Monsanto accountant, was knocked to the floor by the explosion. Instinctively he crawled under his desk. Just then the roof collapsed, but the desk gave him refuge. He managed to crawl away from the ruins before the entire Monsanto plant burned.

Mike Mikeska and Henry Baumgartner, both from the Terminal Railway, were in the thick of the fight to stem the fire on the *Grandcamp*. Both were killed by the blast.

Roy Wade, administrative assistant to the chief of the Texas Public Safety Department, was near the scene at the time of the blast. He said that a steel plate from the ship "knocked a hole in the street as big as an automobile. A steel barge, thirty by one hundred feet, was lifted out of the water by the explosion and left high and dry nearly fifty yards from the dock with the twisted remains of fire trucks on top of it. In one demolished building, a number of burned and blasted bodies were found, as well as a little black dog, very weak but alive."

Al Gerson, a teenager, was working near the wharf area at the Bon Ton Cafe, run by his father and mother. When the blast occurred the roof of the building fell in, and Gerson took shelter beneath the counter. When the crash of timbers subsided, Gerson crawled cautiously from his refuge. He heard a weak cry from his mother and went outside to discover his parents pinned beneath the restaurant portico. Gerson, along with a man who saw what happened, extricated them from the wreckage. Gerson's father was dead, but his injured mother lived.

In those terrible first minutes of disaster Texas City was virtually helpless. Electric facilities and the water system were knocked out of commission and there was no way to halt the spread of the fire. It was the chain reaction that doomed the city. Fiery pieces from the ship hit buildings and started new fires. Like meteors they smashed into gasoline and oil storage tanks, and as these tanks exploded they ignited others. Chemicals from Monsanto were turned into flaming rivers. Within a short time two miles of waterfront were ablaze and the fire was moving inland. Small explosions set off by the fire continued to plague the city.

By mid afternoon help began to arrive from surrounding areas. Doctors, nurses and medical supplies converged at the scene of the disaster. Volunteers poured in from nearby Galveston. Staff members of the American Red Cross flew in from St. Louis and Washington, D.C. Policemen from Dallas, Fort Worth, Beaumont, San Antonio, Port Arthur, and Houston augmented the Texas City force. Army, navy, air force and coast guard teams arrived, and the Army set up field kitchens to feed rescue workers as well as victims. Gas masks, penicillin, blankets, beds, and blood plasma arrived from California and New England. Medical supplies came in from Halifax, where sympathy with Texas City's plight was probably sharper than anywhere else because the Canadian city remembered the *Mont Blanc*.

The people of Texas City reacted to the sudden tragedy in different ways. The first impulse of many was to flee the city, and these took to the roads, clutching favored belongings. Others stoically refused to desert their homes, standing by to face whatever fate might come their way. But as the extent of the threat to Texas City dawned in the minds of those closest to the disaster, police cars with loudspeakers began to move through the debris-littered streets of the city warning the people to evacuate. Already poisonous nitrogen dioxide fumes were claiming victims, and it was obvious that more explosions would occur.

But rescue workers continued to buck the odds as they tried to clear the dock area of the dead and dying. They remained at their tasks even though they knew that another violent explosion was likely to occur at any moment. That explosion, when it came, could be every bit as disastrous as the blast on the *Grandcamp*.The *High Flyer*, its hold filled with 961 tons of ammonium nitrate fertilizer and a large quantity of sulphur, could go up without warning.

All day long the rescue workers on the dock labored in the smoke-filled atmosphere, and all day long the rest of the city tried vainly to put out fires that seemed to spring up on every block. Meantime, tugs moved in to pull the *High Flyer* out into the bay, but she was jammed against the *Wilson B. Keene* in such a way that she could not be budged.

At 6:00 P.M. with fires still raging throughout the city, the Coast Guard officially announced that the *High Flyer* was on fire. At 7:30 city officials, aware now of what ammonium nitrate could do, urged again that reluctant citizens leave the city. This time most of the people obeyed. The long wavering lines of people leaving their homes were reminiscent of evacuations that had occurred during the recent war.

On board the *High Flyer* sailors attempted in vain to stem

the fire, and at 10:30 all hands were ordered ashore. The tugs trying to free the *High Flyer* worked until midnight, despite smoke, flame, and poisonous gases. At midnight they were ordered to give up the attempt because the *High Flyer* was expected to explode within minutes. Police on the dock began herding rescue workers to safer places.

Among those on the dock was a Catholic priest, Father William A. Roach of nearby St. Mary's Church, and a group of rescue workers under his command. His group continued to drag injured people from smoking ruins as he administered last rites to the dying. They refused to leave.

At 1:11 A.M. the *High Flyer* blew up. Father Roach and his helpers were incinerated in the blast.

This second tremendous explosion, almost as powerful as the one before it, rocked Texas City like a cradle in a windstorm. The only saving grace, this time, was that there were fewer people on the waterfront and therefore fewer casualties. But it was the final straw to many people who had refused until now to abandon their homes. In the darkness of the night more terrified people left the city.

The *High Flyer* almost vanished in the tremendous blast. Then, a few minutes later, a second explosion finished her. Like the *Grandcamp*, she ceased to exist. The *Wilson B. Keene*, jammed alongside, was ripped in half and one portion of it was hurled from the ship, crashing through a warehouse and coming to rest 300 feet from the dock. Pieces of the *High Flyer* struck the few oil tanks that had escaped the *Grandcamp* explosion, setting them afire. A flying fragment neatly sliced off the foot of a Salvation Army worker who was leaving the dock at the behest of officers. An 8,000-pound turbine was propelled 4,000 feet, crashing through a roof. Another barge was lifted 200 feet in the air, landing across the broken and charred remains of several small ships at dockside. A ten-car train was smashed against an embankment as if a giant's hand had swept it from its tracks.

W. H. Boucher, a rescue worker who was still on the dock

when the *High Flyer* blew up, miraculously escaped injury. "But," he said, "the blast blew the buttons off my jacket, and a piece of shrapnel cut the right leg off a man standing right next to me. A Red Cross nurse helped me to fix him a tourniquet. Her head was bleeding but she didn't pay any attention to herself. The man was conscious but he didn't say a word. He didn't even moan. I stayed with him until they brought a stretcher and carried him off."

William Bankhead, another volunteer rescue worker, said, "We were one hundred yards from the *High Flyer* when she blew up. The explosion knocked me down and I rolled under an automobile. When the stuff stopped falling, I looked out and saw a lot of men lying on the ground."

One of the most heroic nurses on the scene was Mrs. Clay Martin. She went on working among the injured in spite of eight ribs broken by a piece of flying metal. "It rained steel out there," she said later. "It was awful. A man near me got an eye knocked out. Another fellow got a foot cut off. I was darned lucky to get out with broken ribs."

The fires in the shattered city continued through the long black night, with intermittent explosions as oil tanks or other vulnerable targets of the blaze blew up. But by the morning of April 17, hope began to rise for the shaken citizens of Texas City. The fires were now being contained and it looked as if the worst was over. By April 22 the last of the fires were extinguished and a shocked silence fell over the weary city.

Sadly, Texas City counted its dead, its injured, its homeless, its orphaned. The statistics were sickening: 561 killed, 3,500 injured, most of them seriously; 2,500 homeless and jobless; 844 dependents (widows, children, parents) left behind, some completely without resources. On the battered waterfront, warehouses, refineries, freight yards, chemical plants and other buildings worth a total of $125,000,000 were almost totally destroyed. In addition, there were millions of dollars of damage to the city itself.

All this could well have spelled the end of Texas City as an industrial port city. Before the last fires were snuffed out local businessmen began debating whether rebuilding was financially worthwhile. Some said it was; others were convinced the city was dead. Edgar M. Queeny, chairman of the board of Monsanto Chemical Company, settled the question in dramatic fashion. He announced that Monsanto would rebuild its ruined plant in Texas City.

The announcement proved to be the catalyst the city needed for rehabilitation. Businessmen went to work doggedly to revitalize the city. What happened was one of the most gratifying recoveries from disaster in history. In only one year the population of Texas City had jumped from 17,500 to 27,000. Seven hundred houses had been repaired, 500 new ones had been built, and 900 more begun. The shattered business enterprises along the waterfront were back in operation, and 217 new buildings had emerged from the rubble. Monsanto, the hardest hit of all, was embarked on a multimillion dollar expansion program. The ship channel had been deepened, and longshoremen were loading and unloading ships as they always had.

Three years later, in 1951, the industrial rehabilitation was complete. The huge Monsanto plant had been rebuilt at a cost of $50,000,000. Other industries had poured almost $30,000,000 into repairs and new buildings. Texas City itself had kept pace in laying out eighteen miles of new streets and erecting 1,400 new houses, $1,500,000 worth of new schools, a new police station, two new fire houses, and a health clinic. The revitalization of Texas City had proceeded in a way that surprised the world, and in 1951 Texas City's Mayor Lee Robinson, forsaking grammar for civic pride, said, "People would tell me that Texas City was done for and I'd always say: Just because you get hit hard in the stomach don't mean you're going to lay down and die."

Indeed, Texas City had pushed forward to a new life. In only one area did it retreat. City fathers realistically decided that never again would ammonium nitrate, cause of the 1947 holocaust, be loaded there.

14

Carnage in the Classroom (1958)

PROBABLY no tragedy so deeply touches the hearts of people as one involving a large number of children. That a new life can be snuffed out in seconds is difficult to accept. Today, in Chicago, there are people who will be forever haunted by the year 1958, when a parochial school on the city's West Side was gutted by fire.

The memory of this particularly horrifying disaster still lives in the minds of those who fought the blaze, those who helped in rescue work, and those who stood in frustration behind police barriers. Some lost a child—a young life full of promise and hope—in the awful holocaust. Some saw their children come through the inferno alive. Some were teachers and nuns and priests who labored, not always successfully but always heroically, in a hell of flames, to bring out children. And some fought the terrible blaze with hoses and ladders and, most of all, with courage.

None came away unaffected. Those appalling moments

when the school was ablaze and the anguished cries of children pierced the air have been imbedded in the minds of those who witnessed the disaster. Even in the memories of the firemen—those professionals who, by the nature of their business, might be expected to become hardened to such tragedies—the horror of the catastrophe still lingers.

Our Lady of the Angels Roman Catholic Parochial School—two buildings, one old, one new—was the focal point of a residential neighborhood made up primarily of single family frame homes and a few two-family dwellings. It was a mixed neighborhood ethnically, consisting of second and third generation Chicagoans of Irish, German and Italian extraction. Almost all of its residents were Roman Catholics.

The fire started in the older building, a red brick structure erected in 1910. Originally this building was shared with Our Lady of the Angels Church, the church occupying the first floor and classrooms the second. But in 1939 a new church was built nearby and the old building was given over to classes. Twelve years later the school was remodeled and modernized, but it retained its old-fashioned ceilings, its wood trim, and a few fire hazards that should have been corrected. In 1953 an adjacent building was erected to accommodate additional pupils. This second building was not involved in the fire.

December 1, 1958, was a bright winter day with temperatures in the upper twenties. It was Monday and a new school week had begun. Approximately fifteen hundred grade-school and one hundred twenty kindergarten children were attending classes in the two buildings. They ranged in age from five to fourteen years. The faculty consisted of nine lay teachers and twenty nuns of the Order of the Sisters of Charity of the Blessed Virgin Mary. The principal of the school was Sister Mary St. Florence; the pastor of the church, the Right Reverend Joseph F. Cassen.

The school day was drawing to its close. Some teachers had already assigned tomorrow's homework, and the papers and books needed to do the work were piled neatly on each child's desk. As always, the children waited expectantly for the three o'clock bell to ring, and teachers and nuns passed the time by telling the children stories as the clocks ticked away the final minutes. Outside, mothers were arriving in cars to pick up their children, creating the usual traffic snarl. Others were approaching the school on foot.

In the old building, fifteen boys had been released early to perform special cleanup chores. It was their job to take wastebaskets full of discarded papers to the boiler room and deposit the trash in a large metal container. One of the boys returned from the boiler room and told his teacher that he had "smelled something funny" at the foot of the stairs that led from the basement to the first and second floors, but the teacher attached no importance to his remark at the time. Later another youngster claimed that he had seen a boy lurking around the boiler room who had no business being there. This last, however, was not an unusual occurrence; it was known that boys often went into the basement to sneak a smoke.

In any case, the small blaze that was to grow into one of the country's worst school fires began in a pile of papers and rubbish at the foot of the stairs. Firemen later speculated that the fire may have smoldered for some time before bursting into a larger flame. Whether or not the fire was started by a carelessly tossed match or cigarette is not known, but later investigations pointed to this explanation as the most logical one.

The timing of the fire was particularly tragic. However long it may have smoldered, it burst into dangerous life at 2:42 P.M., only eighteen minutes before the bell would have dismissed the children for the day.

Once the fire started, there was no chance to control it. It roared up the stairway behind a panic-stricken janitor who ran down the halls shouting, "Call the fire department!" at the top of his lungs. It bypassed the first floor, leaped up the stairwell to the second, and burst into the halls where it licked voraciously at the doors of the classrooms. It spread through the school with such savage speed that well-rehearsed fire drills were forgotten by many pupils and teachers. Such drills—on much of the second floor, anyway—would have been ineffective because the fire blocked the exits of most of the rooms within seconds of its discovery.

The surrounding neighborhood knew of the fire almost as soon as those trapped in the building. Instead of hearing the familiar sound of happy voices, the area's residents were shocked to hear cries of anguish. People in the neighborhood descended on the school, led by terrified mothers who dropped their chores and dashed to the rescue of their children. The fire department, truck sirens wailing, appeared as if by magic. A cordon of police assembled, holding frantic parents back from the blazing school.

Inside the building children were dying. Some who opened their classroom doors were immediately overcome by choking black smoke. Some died at their desks from asphyxiation. Others, cut off from all exits by the roaring fire in the hallways, went to the windows of their classrooms—and here is where tragedy multiplied. Had they waited for firemen to run up ladders and take them out, it is likely that many more would have been saved. Instead, the children began to jump from the windows.

The sight of children leaping from high windows was horrifying. One eyewitness said, "Kids were hanging from windowsills, jumping or falling in groups of three or four at a time. Smoke and fire poured from some of the windows."

Pupils on the first floor were the luckiest. The fire, skipping past the first floor as it roared up the stairwell, gave these youngsters time to escape. Many were able to make normal exits by way of the hallways; others crawled out on window ledges and dropped to the ground unhurt. On the second floor the height from the windowsills to the ground was formidable, but the terrified children jumped anyway. Many were killed as they struck the ground; others were seriously injured.

In one second-floor section of the building, where classrooms were not blocked by flame or smoke, teachers and their pupils ran into the halls. Some managed to reach the door to the lone outside fire escape, which dangled precariously down the back of the building. To their dismay they found it stuck shut. Fortunately, a priest from the church realized what the problem was, climbed up the fire escape, and pulled open the door from the outside. All of the children directed to the fire escape were saved.

Meantime, firemen—aided by priests, nuns, lay teachers and janitors—were making heroic efforts to save as many as possible. Firemen ran ladders up to the second floor and plucked youngsters from the windows. They fought their way past heat, flames, and smoke into classrooms. They found some children alive in the smoke-filled rooms, and either carried them from the building or handed them to other fire fighters on the ladders. One nun, who made three trips into the flaming building to rescue children, said in awe, "I felt untold strength."

In many cases rescuers found youngsters sprawled dead on the floor, often untouched by flames but asphyxiated. Some who had gone into the hallways were already charred and twisted corpses. In one classroom firemen found twenty-four children sitting at their desks or crumpled on the floor. All of them were dead. Books containing homework assignments, never to be completed, were tidily stacked on their desks.

There was evidence that some of the children remembered what they had learned in their fire drills, but the inferno on the second floor made that of little use to them. Many had calmly walked into the halls and headed, like well-trained soldiers, for exits. But the rapid pursuit of the fire caught them before they could leave the building. Others panicked, racing to their deaths in the halls or crawling off into corners to die in the billowing smoke.

In a few classrooms determined teachers were able to maintain order. Mrs. Edna Shanahan, one of the lay teachers, soothed the fears of her charges with her soft warm voice. Quietly she told them to wait until firemen put ladders up to the windows. The pupils responded, sitting silently at their desks, determined to obey their teacher in this moment of crisis. While they waited, the Reverend Charles Hunt, assistant pastor of Our Lady of the Angels Church, helped firemen raise a ladder. In one of the most successful single efforts of the day, the children all were rescued.

Teachers used other ingenious methods to save the youngsters. One led her pupils to a stairway, where they froze in fright. Finding them unable to move of their own accord, she laid the children on the top step and rolled them down to safety. It was a bruising descent, but it saved their lives. Another teacher instructed her pupils to form a human chain by clutching each other's clothing, then led the well-controlled group from the burning school.

Sam Tortorice, a young father who lived near the school, became a hero. His daughter was in a second floor classroom when the fire started. Tortorice rushed to the school and battled his way through the smoke to his daughter's room. It was located on an inside ell of the building, and Tortorice saw a man on the ledge of the adjacent wall. Since the fire apparently had not reached that section of the old building, Tortorice swung six children into the man's arms, the last one his daughter.

Most of the children were unable to give coherent accounts of their experiences, but a few articulate youngsters told piteous stories.

Mary Brock, ten years old, exhibited a cool head as others gave in to panic. "When our door [the classroom door] was opened a gust of smoke blew in," she said. "Sister Mary Clara Theresa said, 'Get out of the window, get on the ledge and stay there.' I got out of the window and stood on the ledge. Lots of others jumped."

Mary remained on the ledge until firemen rescued her, but those who jumped were killed on the pavement below. Sister Mary Clara Theresa died in the fire.

Another child who kept her head was Andrea Gagliarco, 12, who stayed at a window calling for help. "Some of the boys jumped out of the window," she related, "but when we looked down we saw them lying there on the ground. We stayed at the window and the firemen saved us."

Linda Barleto, twelve years old, was hurled from a first floor window. She waited on the ledge, afraid to jump, until the heat at her back became unbearable. "Our backs were burning," she said, "and then someone pushed me." Linda was rushed to a hospital. She suffered burns and bruises.

Patricia Perryman, age fourteen, tried to descend from a second floor window on a firemen's ladder. She slipped and slid head-first down the ladder, clutching at the rungs as she fell, and was taken to the hospital with a lacerated arm. She said that fifty-five girls were in her room taking a reading test. "Most of them were frightened." she said, "and some of them started jumping up and running around." She went on to say that a nun finally calmed the children and they waited patiently until a ladder appeared at the window. Patricia was one of the first to descend. "There were some coming out after me," she said. "I don't know if all of them got out."

Ten-year-old Carlos Lozano injured his leg in a frantic leap from a second story window. "Everybody was jumping," he related. "The smoke was terrible. Everybody was screaming. Everybody was trying to get on the fireman's ladders at the same time."

One eight-year-old girl, Concella Ballino, was treated for burns on her hands. Childlike, she failed to grasp the immensity of the tragedy she had survived. She was most worried about a red slipper. "I lost it coming down the ladder," she said plaintively.

Joseph Brocato, eleven years old, was one of the fortunate ones who escaped quickly from the bulding."I was carrying a waste paper basket to the boiler room," he said. "I saw the janitor running from the boiler room. He shouted, 'Call the fire department!' I heard an explosion and then there were flames. My classmate and I ran upstairs and we were told by one of the nuns to go to the church. A lot of children were in the church. We were then told to go home." (This child's reference to an explosion is interesting. There was an early report of a blast that had apparently occurred in the boiler room of the school. But it was later discounted by Fire Commissioner Robert Quinn, who said the boiler room was intact.)

Mrs. Barbara Glowacki, owner of a small grocery store a half block from the school, was electrified by the screams of children and rushed to the scene of the fire. She grabbed many of the youngsters as they fled from the school. "I lined them up nearby," she said. "All the time I was thinking of my own daughter, Helena. But she got out okay. Some of the kids coming out of the school had their hair on fire. I poured water on a number of these."

James Raymond, the chief janitor for the school, was approaching the school from the church when he saw smoke gushing from the windows. He rushed to the building and attempted to lower the fire escape ladder to the ground. But the swing ladder was apparently jammed and would not come down. Before he could get help, the ladder descended and struck him on the head. Raymond was taken to the hospital. (Although no further explanation was given of this incident, it is likely that the priest who ascended the fire escape and opened the door on the second floor may have been coming down the ladder with his charges at the time.)

In the midst of all the confusion and horror, overwrought parents were battling to get through the police and fire lines to rescue their children. One distraught mother carried a red quilted jacket to warm her daughter on the way home. "But I can't find her!" she cried. "I can't find her!" Another woman made fumbling efforts to console her, but she could not concentrate on the task. She, too, had lost her child.

In time the flames were extinguished and firemen were deployed through the gutted building to bring out the dead. The scene was one of horror. Firemen carried out charred and blackened bodies as mothers wailed in agony behind police lines. Later, at the Cook County morgue, grieving parents looked at long rows of small bodies, trying to identify their missing children. In many cases the children could be identified only by their possessions—a pocket knife in a boy's trousers, a ring on a girl's finger, a necklace, a wristwatch. Near the bodies was a heart breaking assortment of belongings gathered at the fire scene—jewelry, pens and pencils, books, trinkets of all sorts, a Mickey Mouse watch. Weeping parents moved past the bodies, stopping in shock and disbelief as they came upon their own child, pointing and saying, "That's Johnny" or "That's little Mary."

The tragedy opened the hearts of Chicagoans. People of all creeds and all walks of life gave money, blood, and skin for grafting. Nurses called hospitals and offered to contribute their time. Professional and businessmen gave their personal services and money. One highly placed businessman said, "I'll dig through rubble, run errands, do anything!" So many people wanted to contribute money to families who had lost a child that Mayor Richard J. Daley asked the First National Bank of Chicago to set up a trust fund to handle the donations.

The Chicago Newspaper Publishers Association contributed $10,000 to the cause; the Allied Florist Association donated sprays of pink and white carnations for the coffins of the children; manufacturers of burial vaults offered them free. Mayor Daley, himself a Roman Catholic layman, asked citizens to pray for the children and their families and ordered all flags in the city to be flown at half-mast until the burial of all the dead.

On the Friday following the disaster a single requiem mass was held by Archbishop Albert Meyer at a nearby armory that accommodated five thousand persons on the drill floor. The Archbishop also performed a mass for the three nuns who had died. The final death toll was sickening—93 in all.

The holocaust at Our Lady of the Angels was the most terrible school fire in Chicago history. Only two school fires in one hundred years had cost more lives than this one. One was an explosion and fire in New London, Texas, on March 18, 1937, which took 294 lives; the other was a fire in Collinwood, Ohio, on March 4, 1908, that killed 176.

The investigation of the fire was promptly launched. Police, firemen and other officials of the city, combined with the office of the Coroner of Cook County and the local office of the Federal Bureau of Investigation, held lengthy hearings. It was finally agreed that the fire had started at the stairwell in the basement among some newspapers and other rubbish piled there. There seemed no other explanation of the fire's origin than that a cigarette or match had started the blaze. It was also decided that a closed steel door leading from the stairwell to the first floor classrooms was instrumental in holding back the fire at that point. But it sent the fire racing up to the second floor where no such door blocked it. Thus, most of those on the first floor escaped while those on the second bore the brunt of the flames and smoke.

Fire Commissioner Quinn testified that the school building had been inspected by the fire department two months earlier—during the celebration of Fire Prevention Week. At that time the single outside fire escape and six other interior exits from the second floor were considered adequate for quick egress.

"All the laws were complied with," Quinn said. "The building actually was what one would term very clean."

Although the building technically conformed to Chicago's fire codes covering old buildings, it was admitted by a city building inspector that schools erected after 1949 were much safer. Despite the fact that Our Lady of the Angels School had been modernized in 1951, it was listed on city records as a "pre-ordinace" building, unaffected by a new building code established in 1949. The 1949 code, among other things, required that stairwells be enclosed with fire-resistant doors and material *at each floor,* a precaution that would have saved numerous lives had Our Lady of the Angels School been so equipped.

The huge fire that swept through the school merely reemphasized that man's mistakes cause most of his troubles. Had more foresight been exhibited by city building inspectors, fire officials and school administrators, this killer-fire could have been either prevented or at least more quickly controlled—and the ghastly toll of 90 children and three adults could have been reduced or eliminated entirely.

Man fears nothing so much as fire. It can turn on him in a moment and sweep away his life. The smallest flame, spreading quickly, engulfs whole cities. And wherever people are crowded together—on shipboard, in theaters, factories, hotels—fire raises again the dread spectres of panic and death.

Here are the great holocausts of the last hundred years—fourteen unforgettable tragedies recounted in vivid detail:

- the flying embers, crumbling walls, and hysterical throngs of the Great Chicago Fire of 1871 that leaves hundreds dead and thousands homeless
- the spectacular close of the era of giant airships, with the flaming end of the *Hindenburg*
- terrifying destruction in a no-exit loft factory in New York City
- the explosion of a munitions ship that nearly consumes the city of Halifax
- a monstrous earthquake and fire that make flourishing San Francisco a crucible of horror